AS/A-LEVEL YEAR 1

STUDENT GUIDE

EDEXCEL

Biology B

Topics 1 and 2

Biological molecules

Cells, viruses and reproduction of living things

Mary Jones

PHILIP ALLAN FOR
HODDER
EDUCATION
AN HACHETTE UK COMPANY

Philip Allan, an imprint of Hodder Education, an Hachette UK company, Blenheim Court, George Street, Banbury, Oxfordshire OX16 5BH

Orders

Bookpoint Ltd, 130 Milton Park, Abingdon, Oxfordshire OX14 4SB

tel: 01235 827827

fax: 01235 400401

e-mail: education@bookpoint.co.uk

Lines are open 9.00 a.m.–5.00 p.m., Monday to Saturday, with a 24-hour message answering service. You can also order through the Hodder Education website: www.hoddereducation.co.uk

© Mary Jones 2015

ISBN 978-1-4718-4384-6

First printed 2015

Impression number 5 4 3 2 1

Year 2018 2017 2016 2015

This guide has been written specifically to support students preparing for the Edexcel AS and A-level Biology B (Topics 1 and 2) examinations. The content has been neither approved nor endorsed by Edexcel and remains the sole responsibility of the author.

Cover photo: Elena Pankova/Fotolia

Typeset by Greenhill Wood Studios

Printed in Italy

Hachette UK's policy is to use papers that are natural, renewable and recyclable products and made from wood grown in sustainable forests. The logging and manufacturing processes are expected to conform to the environmental regulations of the country of origin.

Contents

Content Guidance

Questions & Answers

■ Getting the most from this book

Exam-style questions

Commentary on the questions

Tips on what you need to do to gain full marks, indicated by the icon (e).

Sample student answers

Practise the questions, then look at the student answers that follow.

Commentary on sample student answers

Find out how many marks each answer would be awarded in the exam and then read the comments (preceded by the icon (e)) following each student answer. These indicate exactly how and where marks are gained or lost.

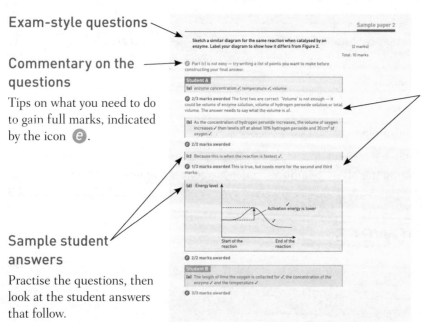

Sample paper 2

Sketch a similar diagram for the same reaction when catalysed by an enzyme. Label your diagram to show how it differs from Figure 2. (2 marks)

Total: 10 marks

(e) Part (c) is not easy — try writing a list of points you want to make before constructing your final answer.

Student A

(a) enzyme concentration ✓, temperature ✓, volume

(e) 2/3 marks awarded The first two are correct. 'Volume' is not enough — it could be volume of enzyme solution, volume of hydrogen peroxide solution or total volume. The answer needs to say what the volume is of.

(b) As the concentration of hydrogen peroxide increases, the volume of oxygen increases ✓ then levels off at about 10% hydrogen peroxide and 30 cm³ of oxygen ✓.

(e) 2/2 marks awarded

(c) Because this is when the reaction is fastest ✓.

(e) 1/3 marks awarded This is true, but needs more for the second and third marks.

(d) Energy level

Activation energy is lower

Start of the reaction   End of the reaction

(e) 2/2 marks awarded

Student B

(a) The length of time the oxygen is collected for ✓, the concentration of the enzyme ✓ and the temperature ✓.

(e) 3/3 marks awarded

Biological molecules • Cells, viruses and reproduction of living things 75

■ About this book

This book is the first in a series of four covering the Edexcel AS and A-level Biology B specifications. It covers Topics 1 and 2:

- Biological molecules
- Cells, viruses and reproduction of living things

This guide has two main sections:

- The **Content Guidance** provides a summary of the facts and concepts that you need to know for these two topics.
- The **Questions & Answers** section contains two specimen papers for you to try, each worth 80 marks. There are also two sets of answers for each question, one from a candidate who is likely to get a C grade and another from a candidate who is likely to get an A grade.

The specification

It is a good idea to have your own copy of the Edexcel Biology B specification. It is you who is going to take this examination, not your teacher, so it is your responsibility to make sure you know as much about the exam as possible. You can download a copy free from www.edexcel.com.

The AS examination is made up of two papers:

- **Paper 1** Core Cellular Biology and Microbiology (1 hour 30 minutes, 80 marks)
- **Paper 2** Core Physiology and Ecology (1 hour 30 minutes, 80 marks)

The A-level examination is made up of three papers:

- **Paper 1** Advanced Biochemistry, Microbiology and Genetics (1 hour 45 minutes, 90 marks)
- **Paper 2** Advanced Physiology, Evolution and Ecology (1 hour 45 minutes, 90 marks)
- **Paper 3** General and Practical Principles in Biology (2 hours 30 minutes, 120 marks)

This book covers content that will be examined in Paper 1 of the AS examination and Papers 1, 2 and 3 of the A-level examination.

What is assessed?

It is easy to forget that your examination is not just testing what you *know* about biology — it is also testing your *skills*. It is difficult to emphasise just how important these are.

The Edexcel examination tests three different assessment objectives (AOs). The following table gives a breakdown of the proportion of marks awarded to each assessment objective in the AS and A-level examinations.

Assessment objective	Outline of what is tested	Percentage of marks (AS)	Percentage of marks (A-level)
AO1	Demonstrate knowledge and understanding of scientific ideas, processes, techniques and procedures	35–37	31–33
AO2	Apply knowledge and understanding of scientific ideas, processes, techniques and procedures: ■ in a theoretical context ■ in a practical context ■ when handling qualitative data ■ when handling quantitative data	41–43	41–43
AO3	Analyse, interpret and evaluate scientific information, ideas and evidence, including in relation to issues, to: ■ make judgements and reach conclusions ■ develop and refine practical design and procedures	20–23	25–27

AO1 is about remembering and understanding all the biological facts and concepts you have covered. AO2 is about being able to *use* these facts and concepts in new situations. The examination paper will include questions that contain unfamiliar contexts or sets of data, which you will need to interpret in the light of the biological knowledge you have. When you are revising, it is important that you try to develop your ability to do this, as well as learning the facts.

AO3 is about practical and experimental biology. A science subject such as biology is not just a body of knowledge. Our knowledge and understanding of biology continues to develop, as scientists find out new information through their research. Sometimes new research means that we have to change our ideas.

You need to develop your skills at doing experiments to test hypotheses, and analysing the results to determine whether the hypothesis is supported or disproved. You need to appreciate why science does not always give us clear answers to the questions we ask. Finally, you will be asked to make judgements and reach conclusions, and need to be able to design and improve experiments and procedures, to produce results we can trust.

Scientific language

Throughout your biology course, and especially in your examination, it is important to use clear and correct biological language. Scientists take great care to use language precisely. If doctors or researchers do not use exactly the correct word when communicating with someone, then what they are saying could be misinterpreted.

Biology has a huge number of specialist terms and it is important that you learn them and use them. Your everyday conversational language, or what you read in the newspaper or hear on the radio, is often not the kind of language required in a biology exam. Be precise and careful in what you write, so that an examiner cannot possibly misunderstand you.

The examination

Time

In all of the examinations, the mark allocation works out at around 1 minute per mark. When you are trying out a test question, time yourself. Are you working too fast? Or are you taking too long? Get used to what it feels like to work at around a mark-a-minute rate.

It is not a bad idea to spend one of those minutes just skimming through the exam paper before you start writing. Maybe one of the questions looks as though it's going to need a bit more of your time than the others. If so, make sure you leave a little extra time for it.

Read the question carefully

This sounds obvious but students lose large numbers of marks by not doing so.

■ There is often vital information at the start that you will need in order to answer the questions themselves. Don't just jump straight to the first set of answer lines and start writing. Start reading at the beginning! Examiners are usually careful not to give you unnecessary information, so if it is there it is probably needed. You may like to use a highlighter to pick out any particularly important bits of information in the question.

■ Look carefully at the command words (the ones right at the start of the question) and do what they say. For example, if you are asked to *explain* something then you won't get many marks — perhaps none at all — if you *describe* it instead. You can find all these words in an appendix near the end of the specification document.

Depth and length of answer

The examiners will give you two useful guidelines about how much you need to write.

■ **The number of marks**. Obviously, the more marks allocated, the more information you need to give. If there are 2 marks, then you will need to give two different pieces of information to get both of them. If there are 5 marks, you will need to write much more.

■ **The number of lines**. This is not such a useful guideline as the number of marks, but it can still help guide you to how much to write. If you find your answer won't fit on the lines, then you probably haven't focused sharply enough on the question. The best answers are short and precise.

Mathematical skills

Like all of the sciences, biology uses mathematics extensively. The specification contains an appendix that lists and describes the mathematical techniques that you will need to be familiar with. You will probably have met most of these before, but make sure that you are confident with all of them. If there are any of which you are uncertain, then do your best to improve your skills in them early on in the course — don't leave it until the last minute, just before the exam. The more you practise your maths skills, the more relaxed you will be about them in the exam.

Content Guidance

■ Topic 1 Biological molecules

Carbohydrates

Carbohydrates are substances whose molecules contain carbon, hydrogen and oxygen atoms, and in which there are approximately twice as many hydrogen atoms as carbon or oxygen atoms.

Monosaccharides and disaccharides

The simplest carbohydrates are monosaccharides. These are sugars. They include glucose (Figure 1), fructose, galactose and ribose. The first three of these each have six carbon atoms, so they are also known as **hexose** sugars. Their formula is $C_6H_{12}O_6$. Ribose (Figure 2) has five carbon atoms, and is a **pentose** sugar.

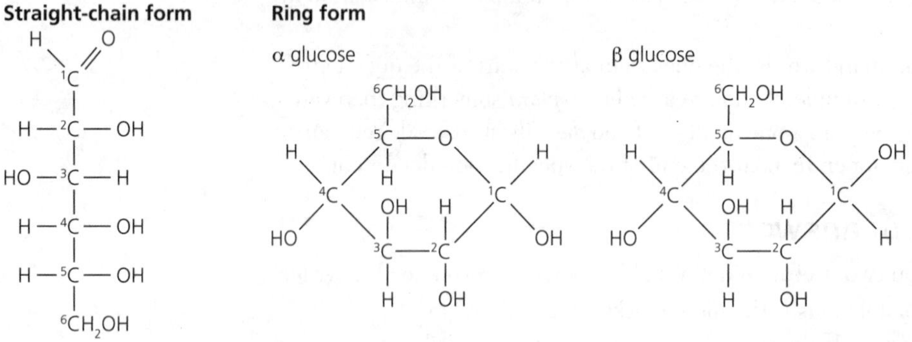

Figure 1 Glucose molecules

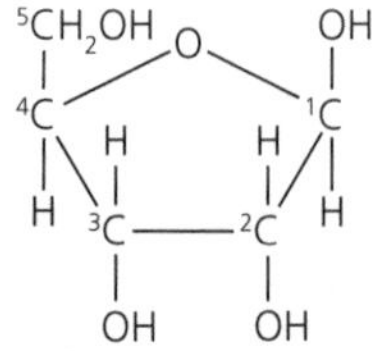

Figure 2 A ribose molecule

Knowledge check 1

What is the difference between the two forms of glucose, alpha and beta, shown in Figure 1?

Knowledge check 2

What is the molecular formula for ribose?

Two monosaccharides can link together to form a disaccharide. For example, two glucose molecules can link to produce maltose. The bond that joins them together is called a **glycosidic bond**. As the two monosaccharides react and the glycosidic bond forms, a molecule of water is released. This type of reaction is known as a **condensation reaction** (Figure 3). Different disaccharides can be formed by linking different monosaccharides (Table 1).

Disaccharide	Monosaccharides that it contains
Maltose	Glucose + glucose
Lactose	Glucose + galactose
Sucrose	Glucose + fructose

Table 1 Disaccharides and the monosaccharides they contain

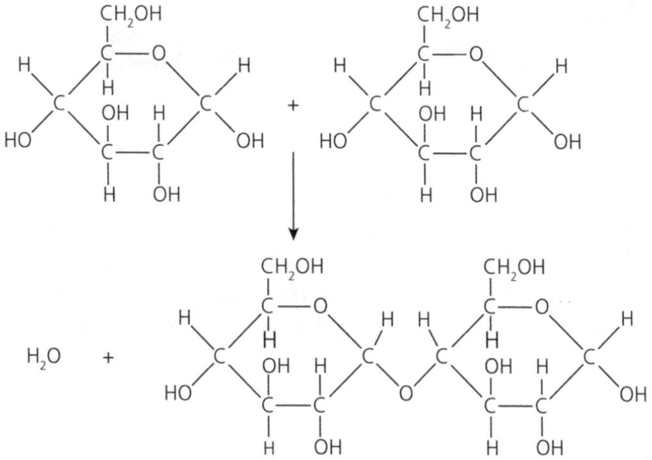

Figure 3 Formation of a disaccharide by a condensation reaction

> **Knowledge check 3**
>
> Using the numbers in Figure 1, state the numbers of the carbon atoms involved in this glycosidic bond.

Disaccharides can be split apart into two monosaccharides by breaking the glycosidic bond. To do this, a molecule of water is added. This is called a **hydrolysis reaction** (Figure 4).

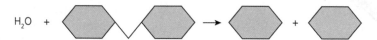

Figure 4 Breakdown of a disaccharide by a hydrolysis reaction

Monosaccharides and disaccharides are good sources of energy in living organisms. They can be used in respiration, in which the energy they contain is used to make ATP. Because they are soluble, they are the form in which carbohydrates are transported through an organism's body. In animals, glucose is transported dissolved in blood plasma. In plants, sucrose is transported in phloem sap.

Polysaccharides

Polysaccharides are substances whose molecules contain hundreds or thousands of monosaccharides linked together into long chains. The monosaccharides are **monomers** that link together to form the polysaccharide **polymers**. Because polysaccharide molecules are so enormous, they do not dissolve in water. This makes them good for storing energy. When needed, they can be hydrolysed to form monosaccharides, which can be used in respiration.

> **Exam tip**
>
> When you are drawing carbohydrate (or any other) molecules, check that you have four bonds on each carbon atom, two bonds on each oxygen atom and one bond on each hydrogen atom.

In animals and fungi, the storage polysaccharide is **glycogen**. It is made of glucose molecules linked together (Figure 5). Most of the glycosidic bonds are between carbon 1 on one glucose, and carbon 4 on the next, so they are called 1–4 links. There are also some 1–6 links, which form branches in the chain.

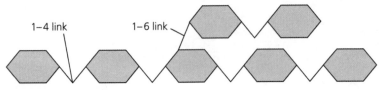

Figure 5 A small part of a glycogen molecule

In plants, the storage polysaccharide is **starch**. Starch is a mixture of two substances, amylose and amylopectin. An amylose molecule is a very long chain of glucose molecules with 1–4 links (Figure 6). It coils up into a spiral, making it very compact. The spiral is held in shape by **hydrogen bonds** between small charges on some of the hydrogen and oxygen atoms in the glucose units. An amylopectin molecule is similar to glycogen.

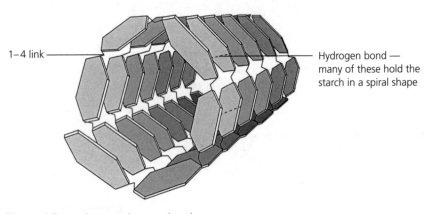

Figure 6 Part of an amylose molecule

Summary

After studying this topic, you should be able to:
- explain the difference between monosaccharides, disaccharides and polysaccharides
- describe the structures of the hexose sugar glucose (alpha and beta forms) and the pentose sugar ribose
- describe how monosaccharides join to form disaccharides by condensation reactions, forming glycosidic bonds
- describe how disaccharides can be split by hydrolysis reactions
- state the monosaccharides involved in the formation of maltose, lactose and sucrose
- describe the structures of the polysaccharides amylose and glycogen, and explain how these are related to their roles in providing and storing energy

Exam tip

Take care with the spellings of glycogen and amylose, and make sure your letters are written clearly. If they look like glucagon or amylase, you will not get credit for your answers.

Knowledge check 4

How does glycogen differ from starch?

Lipids

Lipids, like carbohydrates, also contain carbon, hydrogen and oxygen, but there is a much smaller proportion of oxygen.

Triglycerides

Triglycerides are an important group of lipids. Their molecules are made of a 'backbone' of **glycerol**, to which three fatty acids are attached by **ester bonds** (Figure 7). All lipids are insoluble in water.

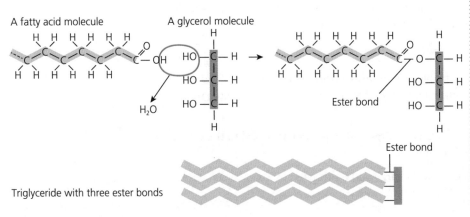

Figure 7 The formation of a triglyceride molecule

Fatty acids have long chains made of carbon and hydrogen atoms. Each carbon atom has four bonds. Usually, two of these bonds are attached to other carbon atoms, and the other two to hydrogen atoms. In some cases, however, there may be only one hydrogen atom attached. This leaves a 'spare' bond, which attaches to the next-door carbon atom, forming a **double bond**. Fatty acids with one or more carbon–carbon double bonds are called **unsaturated** fatty acids, because they do not contain quite as much hydrogen as they could. Fatty acids with no double bonds are called **saturated** fatty acids (Figure 8).

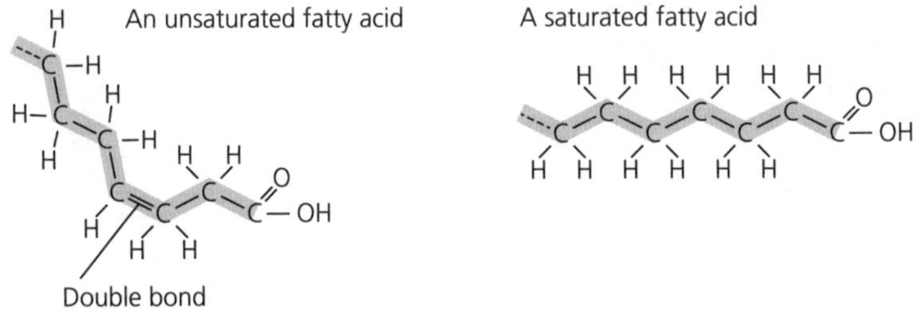

An unsaturated fatty acid

A saturated fatty acid

Double bond

Figure 8 Unsaturated and saturated fatty acids

Lipids containing unsaturated fatty acids are called unsaturated lipids, and those containing completely saturated fatty acids are called saturated lipids. Animal lipids are often saturated lipids. They tend to be fairly solid at room temperature. Plant lipids are often unsaturated, and they tend to be oils, i.e. they are liquid at room temperature.

Knowledge check 5

Look at Figure 7. How is the formation of an ester bond similar to the formation of a glycosidic bond?

Knowledge check 6

What type of reaction will be involved in the breakage of an ester bond?

The functions of triglycerides include:

■ energy storage. Lipid molecules contain energy that can be released by the reactions of respiration, inside mitochondria. Lipids contain more energy per gram than carbohydrates.

■ waterproofing. Lipid molecules are **hydrophobic** (water-hating) and insoluble because they do not have any charges on their molecules. (See the section Water on pages 29–30 for an explanation of this.) They can therefore form waterproof coverings, for example the cuticle of a plant leaf.

■ insulation. Lipids are good heat insulators. For example, lipid stores beneath the skin (in adipose tissue) provide aquatic mammals such as walruses with insulation against the very cold water in which they live.

Phospholipids

A cell membrane consists of a double layer of **phospholipid molecules**. These are lipids with a backbone of glycerol, two fatty acid chains and a phosphate group (Figure 9).

The fatty acid chains have no electrical charge and so are not attracted to the dipoles of water molecules. They are hydrophobic.

The phosphate group has an electrical charge and is attracted to water molecules. It is **hydrophilic**.

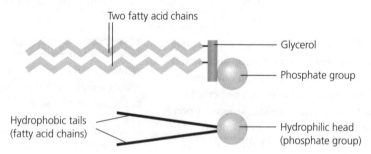

Figure 9 Phospholipids

In water, a group of phospholipid molecules arranges itself into a **bilayer**, with the hydrophilic heads facing outwards into the water and the hydrophobic tails facing inwards, therefore avoiding contact with water (Figure 10).

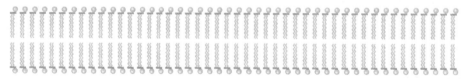

Figure 10 A phospholipid bilayer

This is the basic structure of a cell membrane. The roles of phospholipids in the membranes are:

■ to provide a fluid but well-defined layer in which other molecules, such as proteins and cholesterol, are held

- to prevent the easy movement of non-water-soluble, or large, molecules into or out of the cell

Knowledge check 7

Explain why triglycerides do *not* form a bilayer in water as phospholipids do.

Summary

After studying this topic, you should be able to:

- describe the structure of saturated and unsaturated triglycerides, and explain how ester bonds are formed and broken
- describe how the structure of lipids relates to their roles in energy storage, waterproofing and insulation
- explain how the structure and properties of phospholipids relate to their function in cell membranes

Proteins

Proteins are large molecules made of long chains of amino acids.

Amino acids

All amino acids have the same basic structure, with an amine group and a carboxyl group attached to a central carbon atom (Figure 11). There are 20 different types of amino acid, which differ in the atoms present in the R group. In the simplest amino acid, glycine, the R group is a single hydrogen atom.

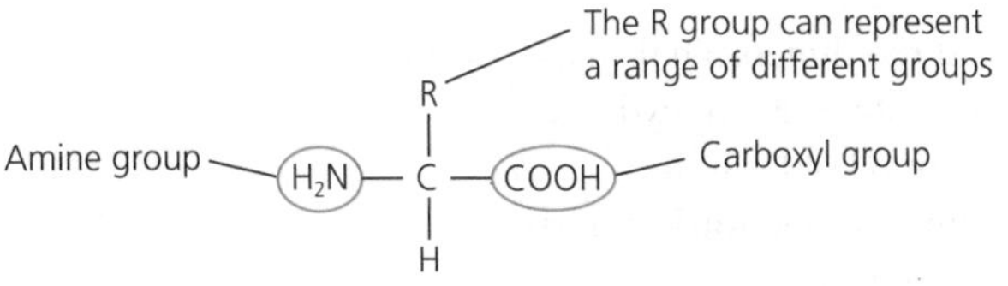

Figure 11 An amino acid

Two amino acids can link together by a condensation reaction to form a **dipeptide**. The bond that links them is called a **peptide bond** (Figure 12). The dipeptide can be broken down in a hydrolysis reaction, which breaks the peptide bond with the addition of a molecule of water.

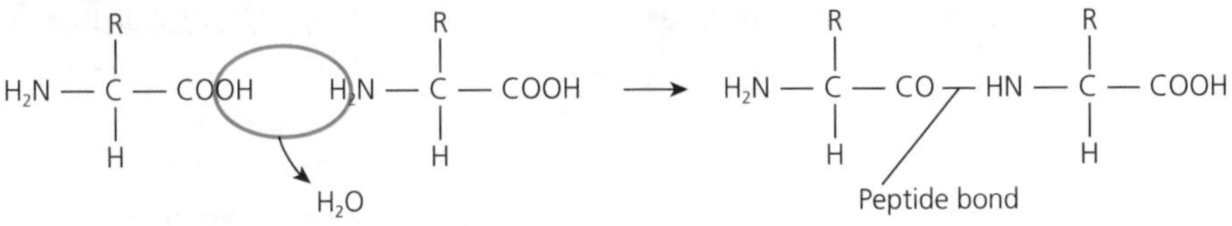

Figure 12 Formation of a dipeptide

Structure of protein molecules

Amino acids can be linked together in any order to form a long chain called a **polypeptide**. A polypeptide can form a protein molecule on its own, or it may associate with other polypeptide molecules to form a protein molecule.

The sequence of amino acids in a polypeptide or protein molecule is called its **primary structure** (Figure 13).

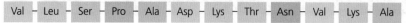

Figure 13 Primary structure

The chain of amino acids often folds or curls up on itself. For example, many polypeptide chains coil into a regular shape called an **alpha helix** (Figure 14). This is held in shape by **hydrogen bonds** between amino acids at different places in the chain. This regular shape is an example of **secondary structure** of a protein.

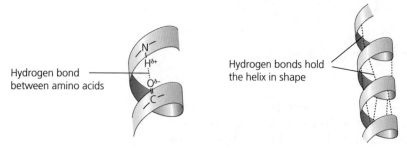

Figure 14 An alpha helix — an example of secondary structure

See page 29 for an explanation of hydrogen bonds.

The polypeptide chain can also fold around on itself to form a three-dimensional shape. This is called the **tertiary structure** of the protein (Figure 15). Once again, hydrogen bonds between amino acids at different points in the chain help to hold it in its particular 3D shape. There are also other bonds involved, including strong **ionic bonds** and **disulfide bonds**. These form between particular R groups on the amino acids.

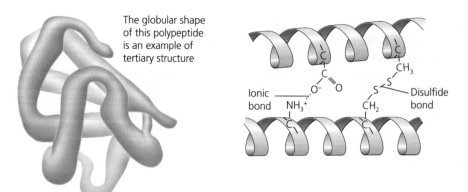

Figure 15 Tertiary structure of a protein

Knowledge check 8

Explain how the tertiary structure of a protein is determined by its primary structure.

The tertiary structure of a protein, and therefore its properties, are ultimately determined by its primary structure.

Some proteins are made of more than one polypeptide chain. The structure formed by the association of these different chains is called the **quaternary structure** of the protein. Haemoglobin is an example of a protein with quaternary structure.

Exam tip

A haemoglobin molecule actually has four polypeptide chains, but this is not why it is said to have quaternary structure. Any protein molecule with two or more polypeptide chains has quaternary structure.

Knowledge check 9

Which types of bond are involved in each of the four levels of protein structure?

Globular and fibrous proteins

Globular proteins have molecules that fold into a roughly spherical three-dimensional shape. Examples include **haemoglobin**, insulin and enzymes. They are often soluble in water and may be physiologically active — that is, they are involved in metabolic reactions within or outside cells. Their solubility is due to having hydrophilic R groups on the outside of the molecule, which readily interact with water.

Haemoglobin has the function of transporting oxygen. In mammals, it is found in solution inside red blood cells. Each of the four polypeptide chains in a haemoglobin molecule contains a haem group, which in turn contains an iron(II) ion. This is able to bond reversibly with oxygen, so haemoglobin can pick up oxygen in the lungs and release it in respiring tissues.

Fibrous proteins have molecules that do not curl up into a ball. They have long, thin molecules, which often lie side by side to form fibres. Examples include keratin (in hair) and **collagen** (in skin and bone). They are not soluble in water and are not generally physiologically active. They often have structural roles. Their insolubility is due to their large size and linear nature, and also having R groups that do not readily interact with water. In contrast to globular proteins, the primary structure of many fibrous proteins has no fixed length, and is made up of many repeating sections of the same kinds of amino acids.

Knowledge check 10

Construct a table to compare the structures of haemoglobin molecules and collagen molecules.

Collagen has the function of providing strength and elasticity in body tissues such as skin, bone and cartilage. Its molecules are made up of three polypeptide chains wound around each other to form a triple helix, which can stretch a little and therefore provide flexibility. Many of these triple helices lie side by side and are cross-linked to form strong fibrils. Fibrils, in turn, lie side by side with other fibrils to form strong fibres.

Summary

After studying this topic, you should be able to:
- describe the structure of an amino acid
- explain how polypeptides and proteins are formed as amino acid monomers are linked by peptide bonds in condensation reactions
- describe the roles of ionic, hydrogen and disulfide bonds
- explain the significance of the primary, secondary, tertiary and quaternary structure of a protein in determining its properties
- describe and explain the roles of haemoglobin and collagen, as examples of globular and fibrous proteins

DNA and protein synthesis

Polynucleotides

Polynucleotides are substances whose molecules are made of long chains of **nucleotides** (Figure 16) linked together. DNA and RNA are polynucleotides.

A nucleotide is made up of:

- a 5-carbon sugar (deoxyribose in DNA; ribose in RNA)
- a phosphate group
- a nitrogen-containing base (adenine, guanine, cytosine or thymine in DNA; adenine, guanine, cytosine or uracil in RNA)

The bases are usually referred to by their first letters, A, G, C, T and U. A and G are **purine** bases, made up of two carbon–nitrogen rings. C, T and U are **pyrimidine** bases, made up of one carbon–nitrogen ring.

Exam tip

Take care not to confuse adenine with adenosine, or thymine with thiamine.

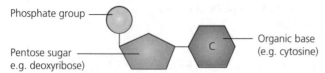

Figure 16 A nucleotide

Nucleotides can link together by the formation of covalent bonds between the phosphate group of one and the sugar of another. This takes place through a condensation reaction. The bonds are called **phosphodiester bonds**. This forms a sugar–phosphate backbone, with bases protruding sideways (Figure 17).

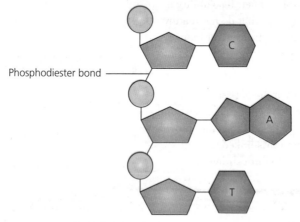

Figure 17 Part of a polynucleotide

A DNA molecule is made up of two polynucleotide strands, held together by hydrogen bonds between the bases on the two strands. The strands run in opposite directions, i.e. they are anti-parallel (Figure 18).

Hydrogen bonding only occurs between A and T and between C and G. This is called **complementary base pairing**.

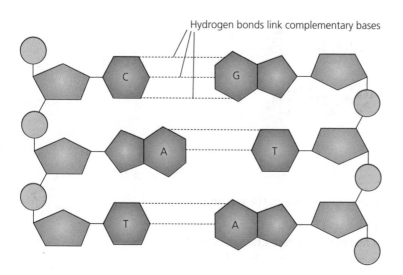

Hydrogen bonds link complementary bases

Figure 18 Part of a DNA molecule

The two strands of nucleotides twist round each other to produce a **double helix**.

DNA replication

New DNA molecules need to be made before a cell can divide. The two daughter cells must each receive a complete set of DNA. The base sequences on the new DNA molecules must be identical with those on the original set. DNA replication (Figure 19) takes place in the nucleus.

- Hydrogen bonds between the bases along part of the two strands are broken by the enzyme **DNA helicase**. This 'unzips' and unwinds part of the molecule, separating the two strands.
- Nucleotides that are present in solution in the nucleus are moving around randomly. By chance, a free nucleotide will bump into a newly exposed one with which it can form hydrogen bonds. Free nucleotides therefore pair up with the nucleotides on each of the DNA strands, always A with T and C with G.
- **DNA polymerase** links together the phosphate and deoxyribose groups of adjacent nucleotides.
- **DNA ligase** is able to link together separate lengths of DNA.

> **Exam tip**
>
> Notice that there are three hydrogen bonds between C and G, and two between A and T.

> **Exam tip**
>
> Do not say that the nucleotides are synthesised at this stage; the nucleotides are already present, and are simply assembled during DNA replication.

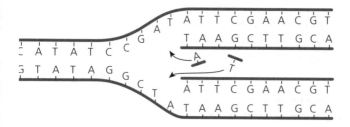

Figure 19 DNA replication

This is called **semi-conservative** replication, because each new DNA molecule has one old strand and one new one (Figure 20).

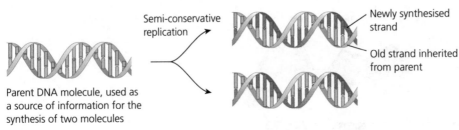

Figure 20 Semi-conservative replication

The genetic code

All of the DNA within a cell, an organism or a species is known as the **genome**.

The sequence of bases in a DNA molecule is a code that determines the sequence in which amino acids are linked together when making a protein molecule. A length of DNA that codes for the sequence of amino acids in one polypeptide, or for one protein, is known as a **gene**.

A series of three bases in a DNA molecule, called a base **triplet**, codes for one amino acid. The triplets do not overlap. For example, Figure 21 shows the sequence of amino acids coded for by a particular length of DNA.

Bases in DNA	T A C	C T G	C A T	C T T
Amino acid in polypeptide	methionine	aspartate	valine	glutamate

Figure 21

There are 20 amino acids. Because there are four bases, there are $4^3 = 64$ different possible combinations of bases in a triplet. Some amino acids therefore are coded for by more than one triplet. For example, the triplets AAA and AAG both code for the amino acid phenylalanine. The code is therefore said to be **degenerate**.

Triplets that code for methionine also act as 'start' signals. Several different triplets act as 'stop' signals.

Many sections of DNA do not code for proteins. Some have as yet unknown functions. Others may be involved in determining whether or not particular genes are used to make proteins.

Protein synthesis

Proteins are made on the **ribosomes** in the cytoplasm, by linking together amino acids through peptide bonds. The sequence in which the amino acids are linked is determined by the sequence of bases on a length of DNA in the nucleus.

Protein synthesis involves RNA:

- Messenger RNA, mRNA, is a single-stranded polynucleotide, not folded. A set of three bases on an mRNA molecule corresponding to a triplet of bases on the DNA molecule is called a **codon**.
- Transfer RNA, tRNA, is also a single-stranded polynucleotide, but it is folded into a clover-leaf shape, held in shape by hydrogen bonds between complementary base pairs (Figure 22). A set of three bases at one end, the **anticodon**, is free to bond

with a complementary codon on mRNA. At the other end is a site where a specific amino acid can bind. A tRNA molecule with a particular anticodon always bonds with a particular amino acid.

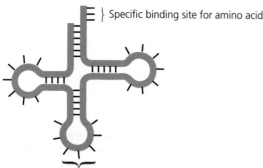

Three bases forming the anticodon

Figure 22 tRNA

Knowledge check 11

Compare the structures of DNA and mRNA.
....

1 In the nucleus, part of a DNA molecule unzips, exposing unpaired bases on both strands.

2 Free RNA nucleotides pair up with the bases on one of the strands, by complementary base pairing (Table 2). The strand with which the RNA nucleotides pair is called the antisense strand. (The other DNA strand is called the sense strand and has the same sequence as the mRNA strand that is produced during translation.)

DNA base in antisense strand	RNA base that pairs with it
A	U
T	A
C	G
G	C

Table 2 Complementary base pairing in protein synthesis

3 The enzyme RNA polymerase links the RNA nucleotides together to produce a messenger RNA (mRNA) molecule.

Steps 1, 2 and 3 are called **transcription**.

4 The mRNA molecule moves out into the cytoplasm and attaches to a ribosome. The ribosome holds the mRNA in a particular position, with two codons inside a groove.

5 tRNA molecules in the cytoplasm become attached to their specific amino acids.

6 One by one, the anticodons of tRNA molecules temporarily pair up with their complementary codons on the mRNA molecule, held in place by the ribosome. This brings their amino acids close together, so that they can be linked by condensation reactions to form peptide bonds.

Knowledge check 12

Explain how tRNA helps to ensure that the 'correct' sequence of amino acids, determined by DNA, is assembled to make a polypeptide.
....

7 The ribosome then moves along the mRNA, so that the next codon is brought into the groove. The next tRNA brings the next amino acid, and so the polypeptide chain gradually lengthens.

Steps 4, 5, 6 and 7 are called **translation**.

Knowledge check 13

Explain the difference between a triplet, a codon and an anticodon.
....

Gene mutation

A gene mutation is a random, unpredictable change in the sequence of bases in a length of DNA coding for a polypeptide or protein.

Mutations are most likely to occur during DNA replication, for example if a 'wrong' base slots into position in the new strand being built. Almost all of these mistakes are immediately repaired by enzymes, but some can persist. Gene mutations that involve a change of a single base are called point mutations.

Gene mutations (Figure 23) can be:

- base deletions. A base is lost from the sequence. This affects the 'meaning' of all the triplets beyond that point.
- base insertions. An extra base is added to the sequence. This also affects the 'meaning' of all the triplets beyond that point.
- base substitutions. One base is replaced by a different base. This only affects one triplet. And even that triplet might still code for the same amino acid, because of the degenerate nature of the genetic code.

Knowledge check 14

Write down the codons on the mRNA molecule produced by transcription of each of the DNA base sequences shown in Figure 23.

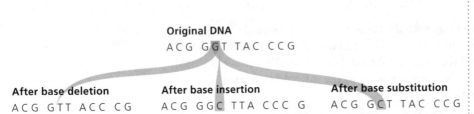

Figure 23 Types of gene mutation

Exam tip

When describing sickle cell anaemia, take care not to confuse haemoglobin molecules with red blood cells. Make quite clear which you are writing about. Use of the word 'it' should be avoided if it is not absolutely clear what 'it' refers to.

Sickle cell anaemia is a condition resulting from a base substitution. The length of DNA coding for two of the polypeptide chains in a haemoglobin molecule normally contains the triplet CTT, but a gene mutation has caused a change to CAT. The normal, CTT, triplet codes for the amino acid glutamic acid, but CAT codes for valine. This change in just one amino acid makes the haemoglobin molecule less able to carry oxygen, less soluble, and more likely to form bonds with other haemoglobin molecules. People whose haemoglobin is of this different form have sickle cell anaemia, in which their red blood cells have a tendency to form sickle shapes (rather than biconcave discs) and block blood capillaries.

Summary

After studying this topic, you should be able to:
- describe the structure of DNA, mRNA and tRNA
- describe how DNA is replicated semi-conservatively, including the roles of the enzymes involved
- explain what is meant by the term *gene*
- describe the genetic code, including triplets coding for amino acids, start and stop triplets (and codons) and its degenerate and non-overlapping nature

- explain how transcription and translation take place, including the roles of mRNA, tRNA and ribosomes
- explain gene mutation (base deletion, insertion and substitution) and describe how gene mutation can result in sickle cell anaemia

Enzymes

An enzyme is a protein that acts as a biological catalyst — that is, it speeds up a metabolic reaction without itself being permanently changed.

The substance present at the start of an enzyme-catalysed reaction is called the substrate, and the new substance (or substances) formed is the product.

Active sites

Enzymes are globular proteins. In one part of the molecule, there is an area called the **active site**, where the substrate molecule can bind. When a substrate molecule attaches to the active site, the shape of the active site changes slightly to make a perfect complementary fit with the substrate. This is called **induced fit**. The need for a complementary shape of active site and substrate means that each enzyme is **specific** for only one substrate.

The R groups of the amino acids at the active site are able to form temporary bonds with the substrate molecule. This pulls the substrate molecule slightly out of shape, causing it to react and form products.

Activation energy

Substrates generally need to be supplied with energy to cause them to change into products. The energy required to do this is called **activation energy**. In a laboratory, you might supply energy by heating to cause two substances to react together.

Enzymes are able to make substances react even at low temperatures. They reduce the activation energy needed to make the reaction take place. They do this by distorting the shape of the substrate molecule when it binds at the enzyme's active site.

Knowledge check 15

High temperatures break hydrogen bonds. Explain how this will affect enzyme activity.

Reactions catalysed by enzymes

Almost every reaction that takes place in an organism's body is catalysed by an enzyme. There are therefore thousands of different enzymes, each catalysing just one type of reaction. Some of these reactions are **intracellular** (take place inside cells) while others are **extracellular** (take place outside cells). Table 3 gives some examples.

Enzyme	Substrate	Product	Examples of where the reaction takes place
Amylase	Starch	Maltose	Extracellularly, in the mouth and duodenum of mammals; also intracellularly in germinating seeds of plants
Catalase	Hydrogen peroxide	Oxygen and water	Intracellularly, in most living cells
DNA polymerase	DNA nucleotides	DNA polynucleotides	Intracellularly, in cell nuclei
ATPase	ATP	ADP and inorganic phosphate	Intracellularly, in all living cells, for example in active transport

Table 3 Some examples of reactions catalysed by enzymes

Factors affecting the rate of enzyme-catalysed reactions

When an enzyme solution is added to a solution of its substrate, the random movements of enzyme and substrate molecules cause them to collide with each other.

As time passes, the quantity of substrate decreases, because it is being changed into product. This decrease in substrate concentration means that the frequency of collisions between enzyme and substrate molecules decreases, so the rate of the reaction gradually slows down. The reaction rate is fastest right at the start of the reaction, when substrate concentration is greatest.

When comparing reaction rates of an enzyme in different circumstances, we should therefore try to measure the initial rate of reaction — that is, the rate of reaction close to the start of the reaction.

Temperature

At low temperatures, enzyme and substrate molecules have little kinetic energy. They move slowly, and so collide infrequently. This means that the rate of reaction is low. If the temperature is increased, then the kinetic energy of the molecules increases. Collision frequency therefore increases, causing an increase in the rate of reaction (Figure 24).

Above a certain temperature, however, hydrogen bonds holding the enzyme molecule in shape begin to break. This causes the tertiary structure of the enzyme to change, an effect called **denaturation**. This affects the shape of its active site. It becomes less likely that the substrate molecule will be able to bind with the enzyme, and the rate of reaction slows down.

The temperature at which an enzyme works most rapidly, just below that at which denaturation begins, is called its optimum temperature. Enzymes in the human body generally have an optimum temperature of about 37 °C, but enzymes from organisms that have evolved to live in much higher or lower temperatures can have much higher or lower optimum temperatures.

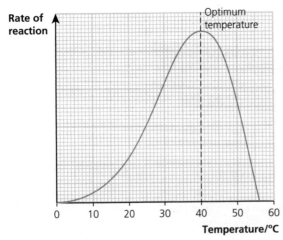

As temperature rises from 0 to about 38°C, an increase in the kinetic energy of the molecules results in increased collision frequency and therefore increased rate of reaction.

As temperature rises above about 40°C, hydrogen bonds in the enzyme break so that it loses its shape and the substrate no longer fits in the active site, resulting in decreased rate of reaction.

Figure 24 How temperature affects the rate of an enzyme-catalysed reaction

Core practical 1a

Investigating the effect of temperature on enzyme activity

You can use almost any enzyme reaction for this, such as the action of catalase on hydrogen peroxide.

Set up several small conical flasks containing the same volume of hydrogen peroxide solution of the same concentration. Stand each one in a water bath at a particular temperature. Use at least five different temperatures over a good range — say between 0 °C and 90 °C. (If time allows, set up three sets of tubes at each temperature. You will then be able to calculate the mean result for each temperature, which will give you a more representative finding.)

Take a set of test tubes and add the same volume of catalase solution to each one. Stand these in the same set of water baths.

Leave all the flasks and tubes to come to the correct temperature. Check with a thermometer.

Take the first flask, dry its base and sides and stand it on a sensitive top-pan balance. Pour in the solution containing catalase that is at the same temperature, and immediately take the balance reading. Record the new balance readings every 30 seconds (or even more frequently if you can manage it) for about 3 minutes. The readings will go down as oxygen is given off.

Repeat with the solutions kept at each of the other temperatures.

Work out the initial rate of each reaction, either taken directly from your readings, or by drawing a graph of mass lost (which is the mass of oxygen) against time for each temperature, and then working out the gradient of the graph over the first 30 seconds or 60 seconds of the reaction (Figure 25).

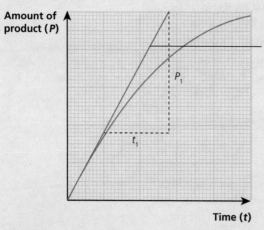

1 Draw a line from the origin through the early part of the graph. Include the part of the line where it is straight. It will only be straight for a short time after the start of the reaction.

2 Calculate the gradient of the line you have drawn.

$$\text{rate} = \frac{P_1}{t_1}$$

Figure 25 Finding the initial rate of reaction

Now you can use your results to plot a graph of initial rate of reaction (y-axis) against temperature.

pH

pH affects ionic bonds that hold protein molecules in shape. Because enzymes are proteins, their molecules are affected by changes in pH. Most enzyme molecules only maintain their correct tertiary structure within a narrow pH range (Figure 26), generally around pH 7. Some, however, require a very different pH; one example is the protein-digesting enzyme pepsin found in the human stomach, which has an optimum pH of 2.

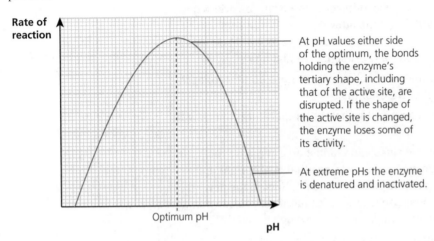

At pH values either side of the optimum, the bonds holding the enzyme's tertiary shape, including that of the active site, are disrupted. If the shape of the active site is changed, the enzyme loses some of its activity.

At extreme pHs the enzyme is denatured and inactivated.

Figure 26 How pH affects the rate of an enzyme-catalysed reaction

Core practical 1b

Investigating the effect of pH on enzyme activity

You can adapt the method described on page 23 for investigating the effect of temperature on the rate of breakdown of hydrogen peroxide by catalase.

Vary pH by using different buffer solutions added to each enzyme solution. (A buffer solution keeps a constant pH, even if acidic or alkaline products are formed.) Keep temperature, enzyme concentration, substrate concentration and total volume of reactants the same for all the tubes. Record, process and display your results as before.

Enzyme concentration

The greater the concentration of enzyme, the more frequent the collisions between enzyme and substrate, and therefore the faster the rate of reaction. However, at very high enzyme concentrations, the concentration of substrate may become a limiting factor, so the rate does not continue to increase if the enzyme concentration is increased (Figure 27).

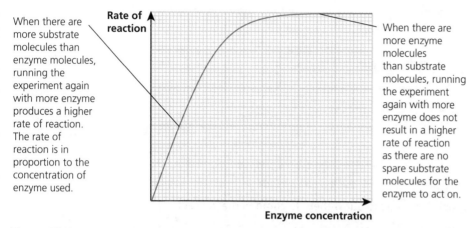

When there are more substrate molecules than enzyme molecules, running the experiment again with more enzyme produces a higher rate of reaction. The rate of reaction is in proportion to the concentration of enzyme used.

When there are more enzyme molecules than substrate molecules, running the experiment again with more enzyme does not result in a higher rate of reaction as there are no spare substrate molecules for the enzyme to act on.

Figure 27 How enzyme concentration affects the rate of an enzyme-catalysed reaction

Core practical 1c

Investigating the effect of enzyme concentration on rate of reaction

You could use the following method to investigate the effect of enzyme concentration on the rate at which the enzyme catalase converts its substrate, hydrogen peroxide, to water and oxygen.

Prepare a catalase solution by liquidising some biological material, such as a handful of celery stalks. This will liberate catalase from the cells. Filter the mixture. The filtrate will contain catalase in solution.

Prepare different dilutions of this solution — for example, see Table 4.

Volume of initial solution/cm^3	Volume of distilled water added/cm^3	Relative concentration of catalase (as a percentage of the concentration of the initial solution)
10	0	100
9	1	90
8	2	80

Table 4

The final solution prepared should be $10\,cm^3$ of distilled water.

Place each solution into a tube fitted with a gas syringe (Figure 28). Use relatively small tubes, so that there is not too much gas in the tube above the liquid, but leave space to add an equal volume of hydrogen peroxide solution at the next step. Ensure that each tube is labelled with a waterproof marker. If time and materials allow, prepare three sets of these solutions.

Place each tube in a water bath at 30 °C.

Take another set of tubes, and add $10\,cm^3$ of hydrogen peroxide solution to each one. The concentration of hydrogen peroxide must be the same in each tube. Stand these tubes in the same water bath.

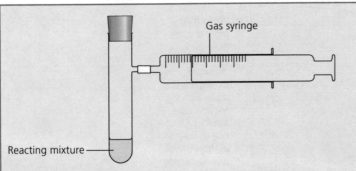

Figure 28 Measuring the rate of gas producing using a gas syringe

Leave all the tubes for at least 5 minutes to allow them to come to the correct temperature.

When ready, add the contents of one of the hydrogen peroxide tubes to the first enzyme tube. Mix thoroughly. Measure the volume of gas collected in the gas syringe after 2 minutes. If you are using three sets, then repeat using the other two tubes containing the same concentration of enzyme.

Do the same for each of the tubes of enzyme. Record the mean volume of gas produced in 2 minutes for each enzyme concentration and plot a line graph to display your results.

Note: if you find that you get measurable volumes of gas sooner than 2 minutes after mixing the enzyme and substrate, then take your readings earlier. The closer to the start of the reaction you make the measurements, the better.

Substrate concentration

The greater the concentration of substrate, the more frequent the collisions between enzyme and substrate, and therefore the faster the rate of the reaction. However, at high substrate concentrations, the concentration of enzyme may become a limiting factor, so the rate does not continue to increase if the substrate concentration is increased (Figure 29).

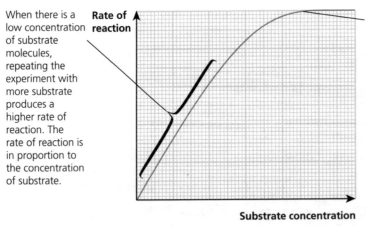

Figure 29 How substrate concentration affects the rate of an enzyme-catalysed reaction

Core practical 1d

Investigating the effect of substrate concentration on the rate of an enzyme-catalysed reaction

You can do this in the same way as described for investigating the effect of enzyme concentration, but this time keep the concentration of catalase the same and vary the concentration of hydrogen peroxide.

Inhibitors

An inhibitor is a substance that slows down the rate at which an enzyme works. Many inhibitors are reversible — that is, they do not attach permanently to the enzyme.

Competitive inhibitors generally have a similar shape to the enzyme's normal substrate. They can fit into the enzyme's active site, preventing the substrate from binding (Figure 30). The greater the proportion of inhibitor to substrate in the mixture, the more likely it is that an inhibitor molecule, and not a substrate molecule, will bump into an active site. The degree to which a competitive inhibitor slows down a reaction is therefore affected by the relative concentrations of the inhibitor and the substrate.

Sometimes the product of the enzyme-catalysed reaction acts as an inhibitor. The greater the concentration of product, the lower the rate of the reaction. This helps to regulate the quantity of product that is formed. This is called **end-product inhibition**.

Non-competitive inhibitors do not have the same shape as the substrate, and they do not bind to the active site. They bind to a different part of the enzyme. This changes the enzyme's shape, including the shape of the active site, so the substrate can no longer bind with it (Figure 30). Even if you add more substrate, it still won't be able to bind, so the degree to which a non-competitive inhibitor slows down a reaction is *not* affected by the relative concentrations of the inhibitor and the substrate.

Knowledge check 16

Temperature, pH, enzyme concentration, substrate concentration, competitive inhibitors and non-competitive inhibitors can all affect the rate of enzyme activity.

a Which of these have their effect by changing the frequency of collisions between the substrate and the enzyme's active site?
b Which have their effect by changing the shape of the active site?

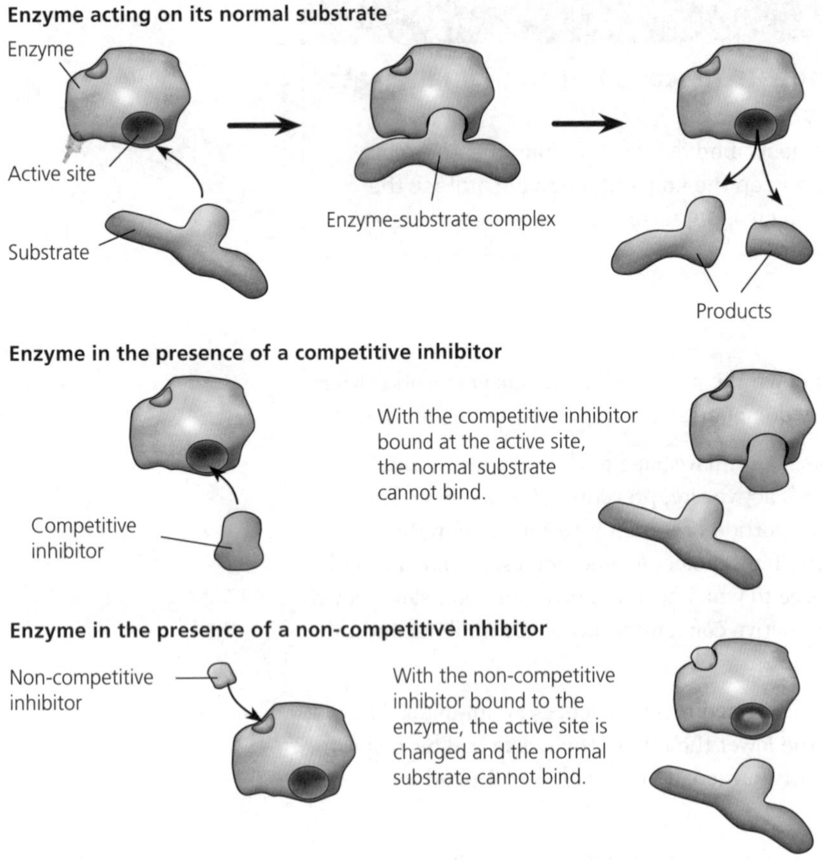

Enzyme acting on its normal substrate

Enzyme

Active site

Substrate

Enzyme-substrate complex

Products

Enzyme in the presence of a competitive inhibitor

With the competitive inhibitor bound at the active site, the normal substrate cannot bind.

Competitive inhibitor

Enzyme in the presence of a non-competitive inhibitor

Non-competitive inhibitor

With the non-competitive inhibitor bound to the enzyme, the active site is changed and the normal substrate cannot bind.

Figure 30 Competitive and non-competitive enzyme inhibitors

Summary

After studying this topic, you should be able to:
- describe enzymes as globular proteins that act as catalysts for a wide range of intracellular and extracellular reactions, by reducing activation energy
- explain the concepts of specificity and the induced fit hypothesis
- explain how temperature, pH, substrate concentration, enzyme concentration, competitive inhibitors and non-competitive inhibitors affect the rate of enzyme activity
- describe how to measure the initial rate of an enzyme-controlled reaction, and explain why this is important
- be able to describe how to carry out investigations into any of these factors on the initial rate of an enzyme-controlled reaction

Inorganic ions

Living organisms require inorganic ions in relatively small quantities. Plants obtain them from the soil, and animals obtain them from the food that they eat. Table 5 summarises the roles of five inorganic ions in plants.

Ion	Symbol	Role in plants
Nitrate	NO_3^-	Formation of amino acids for protein synthesis Formation of nitrogenous bases for production of DNA and RNA
Calcium	Ca^{2+}	Formation of calcium pectate, which makes up the middle lamella between adjacent plant cell walls and holds them together
Magnesium	Mg^{2+}	Formation of chlorophyll, required for the absorption of energy from light for photosynthesis
Phosphate	PO_4^-	Formation of ADP, ATP and nucleotides

Table 5 The role of inorganic ions in plants

Summary

After studying this topic, you should be able to:

- describe the roles of nitrate ions, calcium ions, magnesium ions and phosphate ions in plants

Water

About 80% of the body of an organism is water. Water has unusual properties compared with other substances, because of the structure of its molecules. Each water molecule has a small negative charge (δ^-) on the oxygen atom and a small positive charge (δ^+) on each of the hydrogen atoms. This is called a **dipole**.

There is an attraction between the δ^- and δ^+ parts of neighbouring water molecules. This is called a **hydrogen bond**.

A single water molecule

Hydrogen bonding between water molecules

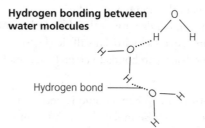

Hydrogen bond

Figure 31 Water molecules

Solvent properties of water

The dipoles on water molecules make water an excellent solvent for any substances that have charges on their molecules. Such substances are said to be **polar**. Most ionic compounds are also soluble in water, because of the charges on their ions. For example, if you stir sodium chloride into water, the sodium and chloride ions separate and spread between the water molecules — they dissolve in the water. This happens because the positive charge on each sodium ion is attracted to the small negative charge on the oxygen of the water molecules. Similarly, the negative chloride ions are attracted to the small positive charge on the hydrogens of the water molecules.

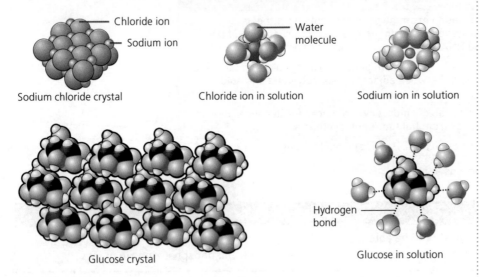

Figure 32 Water as a solvent

Because it is a good solvent, water helps to transport substances around the bodies of organisms. For example, the blood plasma of mammals is mostly water, and carries many substances in solution, including glucose, oxygen and ions such as sodium. Water also acts as a medium in which metabolic reactions can take place, as the reactants are able to dissolve in it.

Thermal properties of water

Water is liquid at normal Earth temperatures. The hydrogen bonds between water molecules prevent them flying apart from each other at normal temperatures on Earth. Between 0 °C and 100 °C water is in the liquid state. The water molecules move randomly, forming transitory hydrogen bonds with each other. Other substances whose molecules have a similar structure, such as hydrogen sulfide (H_2S), are gases at these temperatures, because there are no hydrogen bonds to attract their molecules to each other.

Water has a **high latent heat of evaporation**. When a liquid is heated, its molecules gain kinetic energy, moving faster. Those molecules with the most energy are able to escape from the surface and fly off into the air. A great deal of heat energy has to be added to water molecules before they can do this, because the hydrogen bonds between them have to be broken. When water evaporates, it therefore absorbs a lot of heat from its surroundings. The evaporation of water from the skin of mammals when they sweat therefore has a cooling effect. Transpiration from plant leaves is important in keeping them cool in hot climates.

Water has a **high specific heat capacity**. Specific heat capacity is the amount of heat energy that has to be added to a given mass of a substance to raise its temperature by 1 °C. Temperature is related to the kinetic energy of the molecules — the higher their kinetic energy, the higher the temperature. A lot of heat energy has to be added to water to raise its temperature, because much of the heat energy is used to break the hydrogen bonds between water molecules, not just to increase their speed of movement. This means that bodies of water, such as oceans or lakes, do not change their temperature as easily as air does. It also means that the bodies of organisms, which contain large amounts of water, do not change temperature easily.

Water **freezes from the top down**. Like most substances, liquid water becomes more dense as it cools, because the molecules lose kinetic energy and get closer together. However, when it becomes a solid (freezes), water becomes less dense than it was at 4 °C, because the molecules form a lattice in which they are more widely spaced than in liquid water at 4 °C. Ice therefore floats on water. The layer of ice then acts as an insulator, slowing down the loss of heat from the water beneath it, which tends to remain at 4 °C. The water under the ice therefore remains liquid, allowing organisms to continue to live in it even when air temperatures are below the freezing point of water.

Surface tension and incompressibility

We have seen that water molecules are attracted to each other. Water molecules at the surface of a body of liquid do not have any others above them, so all the attraction is sideways and downwards. This is called **surface tension**. It makes the water behave as though there is a 'skin' at the surface, which is strong enough to support small animals.

In the liquid state, water molecules cannot be pushed any closer together. This means that water cannot be compressed. Some animals make use of this by having hydrostatic skeletons. Earthworms, for example, have fluid-filled compartments in their bodies that help to hold them in shape, and transmit forces from their muscles. Plants rely on the incompressibility of water to enable turgid cells to provide support, for example to hold a leaf out horizontally.

Summary

After studying this topic, you should be able to:
- explain the dipolar nature of water, and how this leads to the formation of hydrogen bonds
- explain why water is able to act as a solvent for small polar molecules and ions
- explain why water has a high latent heat of evaporation, a high specific heat capacity and maximum density at 4 °C, and explain why these properties are important to living organisms
- describe how surface tension and the incompressibility of water are significant for living organisms

■ Topic 2 Cells, viruses and reproduction of living things

Eukaryotic and prokaryotic cells

All living organisms are made up of cells, and a cell can be thought of as the basic unit of living things. There are two basic types of cell — **prokaryotic** cells and **eukaryotic** cells. Prokaryotic cells (Figure 33) are found in bacteria and archaea. Eukaryotic cells (Figures 34 and 35) are found in animals, plants and fungi.

All multicellular organisms are made up of eukaryotic cells. These cells are often specialised for different functions. A group of similar cells carrying out a particular function is called a tissue. Different tissues group together to produce organs, and organs work together as organ systems.

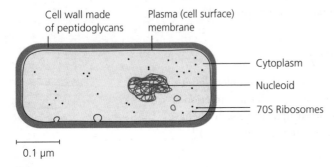

Figure 33 Structure of a prokaryotic cell

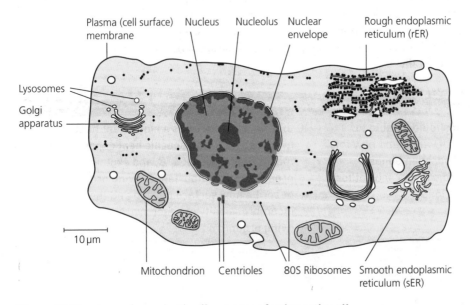

Figure 34 Structure of an animal cell — a type of eukaryotic cell

Knowledge check 17

Approximately how many times longer is the animal cell in Figure 34 than the bacterial cell in Figure 33?

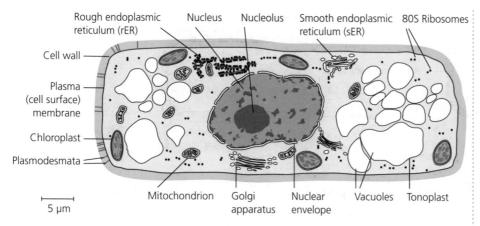

Figure 35 Structure of a plant cell — a type of eukaryotic cell

Knowledge check 18

Construct a table to compare the structure of a prokaryotic cell and a eukaryotic cell.

Bacteria cells can be Gram-positive or Gram-positive (Figure 36). When stained with Gram stain, Gram-positive bacteria take up the stain and look purple, whereas Gram-negative bacteria do not. This is because the cell walls of Gram-positive bacteria have a thick outer layer of peptidoglycan, which absorbs the stain. In Gram-negative bacteria, the cell wall lies between two cell membranes. The outer membrane protects these bacteria from several types of antibiotics, including penicillin.

Functions of organelles

The structures found inside cells are called **organelles**. Table 6 summarises their functions.

Organelle	Function
Nucleus	Contains DNA, providing information for protein synthesis
Nucleolus	Contains DNA, providing information for the formation of RNA for constructing ribosomes
Ribosomes	The site of protein synthesis
Rough endoplasmic reticulum (rER)	The attached ribosomes synthesise proteins; the ER transports proteins to the Golgi body
Smooth endoplasmic reticulum (sER)	Synthesis of lipids and steroids
Mitochondria	The site of the reactions of aerobic respiration, in which ATP is made
Centrioles	Guide the formation of spindle fibres during mitosis and meiosis
Lysosomes	Contain hydrolytic enzymes that digest old organelles, bacteria and so on
Golgi apparatus	Modifies and packages proteins synthesised in the cell
Cell wall	Supports plant cells and prevents them bursting when turgid
Chloroplasts	The site of the reactions of photosynthesis, in which carbon dioxide and water are converted to carbohydrates
Vacuole	In plant cells, contains cell sap, which is an aqueous solution of many different compounds
Tonoplast	The cell membrane surrounding the vacuole, which controls what enters and leaves it

Table 6 Functions of organelles in eukaryotic cells

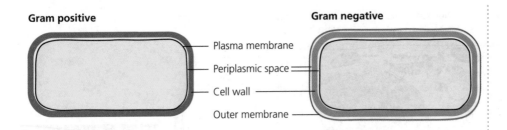

Gram positive   **Gram negative**

- Plasma membrane
- Periplasmic space
- Cell wall
- Outer membrane

Figure 36 Cell walls of Gram-positive and Gram-negative bacteria

Functions of the rER and Golgi apparatus

Proteins that are to be exported from a cell — for example extracellular enzymes — are made on ribosomes attached to the rough endoplasmic reticulum (rER). Amino acids are strung together into a long chain on the ribosome and joined together by peptide bonds. As the chain forms, it is fed through the membrane of the endoplasmic reticulum so that the protein ends up inside the space (cisternum) between the endoplasmic reticulum membranes.

Part of the cisternum, with the protein molecules inside it, breaks off to form a membrane-bound vesicle. This moves towards the Golgi apparatus. The vesicles fuse to the outer (convex) face of the Golgi apparatus.

Inside the Golgi apparatus, the protein molecules are modified, for example by having carbohydrate groups added to them to produce glycoproteins.

Vesicles containing these modified proteins break away from the inner (concave) face of the Golgi apparatus. The vesicles travel to the cell surface membrane, with which they fuse. Their membranes become part of the cell surface membrane and their contents — the proteins — are deposited outside the cell (Figure 37).

> **Exam tip**
> Cells also contain ribosomes not attached to the rER, where non-export proteins are made.

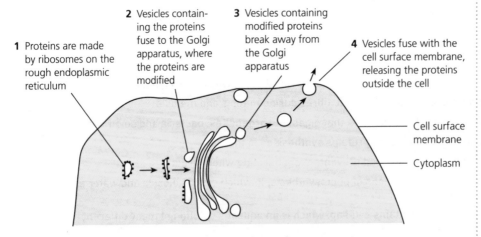

1 Proteins are made by ribosomes on the rough endoplasmic reticulum

2 Vesicles containing the proteins fuse to the Golgi apparatus, where the proteins are modified

3 Vesicles containing modified proteins break away from the Golgi apparatus

4 Vesicles fuse with the cell surface membrane, releasing the proteins outside the cell

- Cell surface membrane
- Cytoplasm

> **Exam tip**
> Note that the membrane of the vesicle is not usually lost from the cell along with its contents — the membrane remains attached to the cell.

Figure 37 Formation and secretion of extracellular enzymes

Table 7 summarises the functions of organelles in prokaryotic cells.

Organelle	Function
Nucleoid	The area of the cell containing DNA; it is not surrounded by a membrane; the DNA contains instructions for protein synthesis
Plasmids	Small circular DNA molecules that contain information for protein synthesis of, for example, substances that provide resistance against antibiotics; plasmids can be transferred from one cell to another
Ribosomes	The site of protein synthesis
Cell wall	Supports the cell and prevents it bursting

Table 7 Functions of organelles in prokaryotic cells

Microscopy

Light microscopes and electron microscopes

Most cells are very small, and a microscope is needed to see their structures. You will use a light microscope during your AS course. Light rays pass through the specimen on a slide and are focused by an objective lens and an eyepiece lens. This produces a magnified image of the specimen on the retina of your eye. Alternatively, the image can be projected onto a screen, or recorded by a camera. Coloured stains can be used to make particular parts of the cell visible. Different stains are taken up by different parts of cells.

An electron microscope uses beams of electrons rather than light rays. The specimen has to be very thin and must be placed in a vacuum, to allow electrons to pass through it. The electrons are focused onto a screen, or onto photographic film, where they form a magnified image of the specimen. Heavy metals, such as gold, do not allow electrons to pass through them, so these can be used as 'stains' to make particular parts of the specimen look darker than others.

Magnification and resolution

Magnification can be defined as:

$$\text{magnification} = \frac{\text{size of image}}{\text{actual size of object}}$$

This can be rearranged to:

$$\text{actual size of object} = \frac{\text{size of image}}{\text{magnification}}$$

There is no limit to the amount you can magnify an image. However, the amount of useful magnification depends on the **resolution** of the microscope. This is the ability of the microscope to distinguish two objects as separate from one another. The smaller the objects that can be distinguished, the higher the resolution.

Resolution is determined by the wavelength of the rays that are being used to view the specimen. The wavelength of a beam of electrons is much smaller than the wavelength of light. An electron microscope can therefore distinguish between much smaller objects than a light microscope — in other words, an electron microscope has a much higher resolution than a light microscope. We can therefore see much more fine detail of a cell using an electron microscope than using a light microscope.

Content Guidance

As cells are very small, we have to use units much smaller than millimetres to measure them. These units are micrometres, µm, and nanometres, nm.

$1\,mm = 1 \times 10^{-3}\,m$
$1\,µm = 1 \times 10^{-6}\,m$
$1\,nm = 1 \times 10^{-9}\,m$

To change mm into µm, multiply by 1000.

Knowledge check 19

How many micrometres are there in 2.3 cm?

Knowledge check 20

Convert 31.5 µm to mm.

Core practical 2

Magnification calculations

You should be able to work out the real size of an object if you are told how much it has been magnified.

For example, the drawing of a mitochondrion in Figure 38 has been magnified 100 000 times.

Figure 38

Use your ruler to measure its length in mm. It is 50 mm long.

As it is a very small object, convert this measurement to µm by multiplying by 1000.

$50 \times 1000 = 50\,000\,µm$

Substitute into the equation:

$$\text{actual size of object} = \frac{\text{size of image}}{\text{magnification}}$$

$$= \frac{50\,000}{100\,000}$$

$$= 0.5\,µm$$

You can also use a scale bar to do a similar calculation for the drawing of a chloroplast in Figure 39.

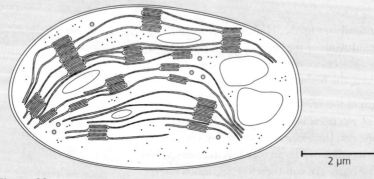

2 µm

Figure 39

continued

Measure the length of the scale bar.

Calculate its magnification using the formula:

$$\text{magnification} = \frac{\text{size of image}}{\text{actual size of object}}$$

$$= \frac{\text{length of scale bar}}{\text{length the scale bar represents}}$$

$$= \frac{20\,000}{2} = \times 10\,000$$

Measure the length of the image of the chloroplast in mm, and convert to µm. You should find that it is 80 000 µm long.

Calculate its real length using the formula:

$$\text{actual size of object} = \frac{\text{size of image}}{\text{magnification}}$$

$$= \frac{80\,000}{10\,000} = 8\,\mu m$$

Measuring cells using a graticule

An eyepiece graticule or eyepiece micrometer is a little scale bar that you can place in the eyepiece of your light microscope. When you look down the microscope, you can see the graticule as well as the specimen.

The graticule is marked off in 'graticule units', so you can use it to measure the specimen you are viewing in these graticule units. Just turn the eyepiece so that the graticule scale lies over the object you want to measure. It will appear as in Figure 40.

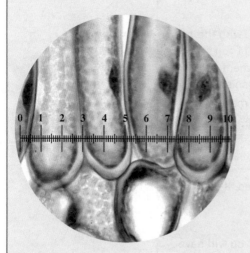

Figure 4

We can say here that the width of one cell is 23 graticule units.

The graticule units have to be converted to real units, such as mm or µm.

continued

This is called calibration. To do this, you use a special slide called a stage micrometer that is marked off in a tiny scale. There should be information on the slide that tells you the units in which it has been marked. The smallest markings are often 0.01 mm apart — that is, 10 μm apart.

Take the specimen off the stage or the microscope and replace it with the stage micrometer. Focus on it using the same objective lens as you used for viewing the specimen.

Line up the micrometer scale and the eyepiece graticule scale. You can do this by turning the eyepiece, and by moving the micrometer on the stage. Make sure that two large markings on each scale are exactly lined up with each other. You should be able to see something like Figure 41.

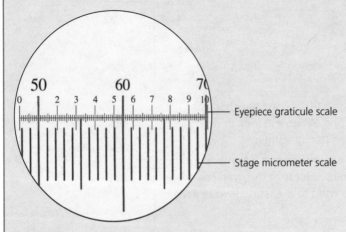

Eyepiece graticule scale

Stage micrometer scale

Figure 41

You can see that the 50 mark on the stage micrometer is lined up with the 1.0 mark on the eyepiece graticule. Work along towards the right until you see another two lines that are exactly lined up. There is a good alignment of 68 on the stage micrometer and 9.0 on the eyepiece graticule. So you can say that:

80 small eyepiece graticule markings = 18 stage micrometer markings

$$= 18 \times 0.01 \text{ mm} = 0.18 \text{ mm} = 180 \text{ μm}$$

So:

1 small eyepiece graticule marking = $\frac{180}{80}$ = 2.25 μm

Now we can calculate the real width of the plant cell we measured. It was 23 eyepiece graticule units long. So its real width is:

23 × 2.25 = 51.75 μm

If you want to look at something using a different objective lens, you will have to do the calibration of eyepiece graticule units all over again using this lens. Once you have done it, you can save your calibrations for the next time you use the same microscope with the same eyepiece graticule and the same objective lens.

Summary

After studying this topic, you should be able to:
- describe the ultrastructure of eukaryotic cells (animal cells and plant cells) and prokaryotic cells
- describe the functions of the organelles in eukaryotic cells and prokaryotic cells
- explain how magnification and resolution can be achieved using light and electron microscopy, including the use of stains
- use a light microscope, eyepiece graticule (eyepiece micrometer) and stage micrometer to measure the size of objects seen through the microscope

Viruses

Viruses are not made of cells. They are not normally considered to be living organisms because they do not display any of the characteristics of life except reproduction, and even this only occurs when they have invaded a living cell.

Classification of viruses

Viruses are classified according to their structure and the type of nucleic acid that they contain (Figure 42). All viruses contain either DNA or RNA. They also contain enzymes, such as proteases, that are used for entering cells or reproducing inside a cell. The DNA is surrounded by a **capsid** made of protein. HIV (the human immunodeficiency virus) has a lipid membrane that is derived from the cell in which it was formed.

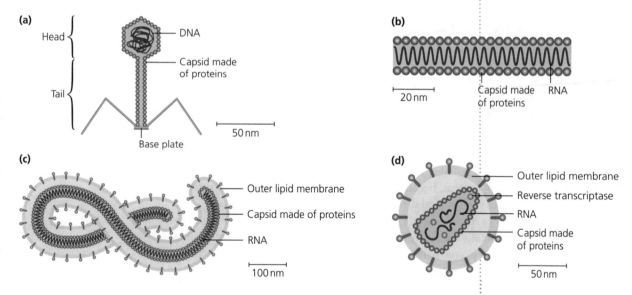

(a)

Head
Tail

DNA
Capsid made of proteins
50 nm
Base plate

(b)

20 nm
Capsid made of proteins   RNA

(c)

Outer lipid membrane
Capsid made of proteins
RNA
100 nm

(d)

Outer lipid membrane
Reverse transcriptase
RNA
Capsid made of proteins
50 nm

Figure 42 Viruses: (a) lambda phage — a DNA virus; (b) tobacco mosaic virus — an RNA virus; (c) ebola virus — an RNA virus; (d) HIV — an RNA retrovirus

Viral reproduction: the lytic cycle

A virus attaches to a receptor on the cell surface membrane of a host cell. Only certain types of cell have receptors to which a particular virus can attach. The virus then injects its DNA or RNA into the cell. In retroviruses, one of the virus's own enzymes, reverse transcriptase, then makes a DNA 'copy' of the viral RNA.

The viral DNA then uses the host cell's enzymes and ribosomes to make new copies of the viral nucleic acid and proteins. These are assembled to make new viruses, which burst out of the cell and destroy it. This is called **lysis**. The viruses then invade other cells nearby and the cycle continues (Figure 43).

> **Exam tip**
>
> Any type of cell can be invaded by a virus. Viruses that attack bacteria are called bacteriophages, or just phages.

1 A virus attaches to a receptor on the host cell.

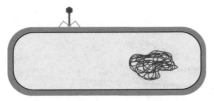

2 The virus injects its RNA or DNA into the cell.

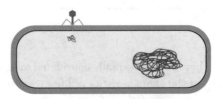

3 The viral DNA uses the host cell's enzymes and ribosomes to copy its nucleic acid and proteins.

4 Lysis — the nucleic acid and proteins are assembled to make new viruses, which burst out of the cell and destroy it.

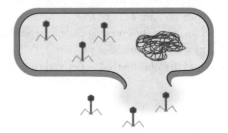

Figure 43 The lytic cycle

Latency

Some viruses, such as HIV, do not make new viruses immediately after invading a host cell. Instead, they insert viral DNA into the host cell chromosomes, where it can lie dormant for many years. No new virus particles are produced during the latent period. Eventually, the viral DNA is activated and expressed, new viruses are formed and the host cell is destroyed.

Viral diseases are difficult to treat, because antibiotics do not affect viruses. Instead, dealing with outbreaks of serious viral diseases, such as the ebola outbreak of 2014–15 in western Africa, focuses on trying to prevent the spread of the virus among populations. Where outbreaks are severe, lethal and spreading rapidly, it may become ethically justifiable to treat people who would otherwise almost certainly die by using recently produced drugs that have not yet been through the full trial procedure.

Summary

After studying this topic, you should be able to:
- describe how viruses are classified according to their structure and type of nucleic acid
- describe the structures of a lambda phage, tobacco mosaic virus and the human immunodeficiency virus
- describe the lytic cycle and latency

Eukaryotic cell cycle and division

Cell division

A multicellular organism begins as a single cell. That cell divides repeatedly to produce all the cells in the adult organism.

Mitosis and the cell cycle

The type of cell division involved in growth is called **mitosis**. Mitosis is also involved in the production of new cells to repair damaged tissue, and in asexual reproduction, in which a single parent gives rise to genetically identical offspring.

As an organism grows, many of its cells go through a continuous cycle of growth and mitotic division called the **cell cycle** (Figure 44). The times at which mitosis occurs are regulated in the body, so that cells only divide as and when new cells are required.

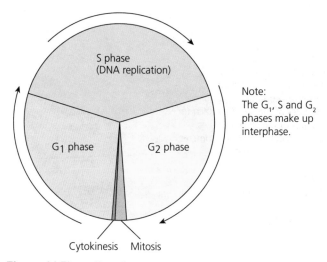

Note:
The G_1, S and G_2 phases make up interphase.

Figure 44 The cell cycle

The cell cycle is made up of three main stages: interphase, mitosis and cytokinesis.

For most of the cell cycle, the cell is in **interphase**. In the G_1 and G_2 phases, the cell continues with its normal activities. It also grows, as the result of the production of new molecules of proteins and other substances, which increase the quantity of cytoplasm in the cell.

DNA replication takes place in the S phase, so that there are two identical copies of each DNA molecule in the nucleus. Each DNA molecule forms one chromosome, so after replication is complete each chromosome is made of two identical DNA molecules. They are called chromatids and remain joined together at a point called the centromere (Figure 45).

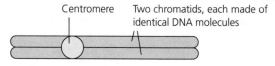

Figure 45 A chromosome before cell division

Exam tip

Note that mitosis does not include interphase or cytokinesis.

Knowledge check 21

Assuming that the pie chart in Figure 44 is drawn to scale, approximately what proportion of the cell cycle involves mitosis?

The cell then enters the second stage of the cell cycle, mitosis. During mitosis, the two chromatids split apart and are moved to opposite ends of the cell. A new nuclear envelope then forms around each group. These two nuclei each contain a complete set of DNA molecules identical to those in the original (parent) cell. Mitosis produces two genetically identical nuclei from one parent nucleus.

After mitosis is complete, the cell enters **cytokinesis**, in which the cell divides into two, with one of the new nuclei in each of the two new cells. These two daughter cells are genetically identical to each other and their parent cell (Figure 46).

Prophase

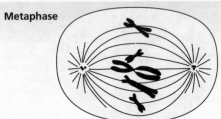

- The chromosomes condense
- The centrioles duplicate
- The centriole pairs move towards each pole
- The spindle begins to form

Metaphase

- The nuclear envelope disappears
- The centriole pairs are at the poles
- The spindle is completely formed
- The chromosomes continue to condense
- The microtubules of the spindle attach to the centromeres of the chromosomes
- The microtubules pull on the centromeres, arranging them on the equator

Anaphase

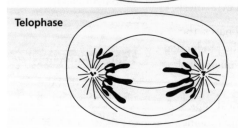

- The links between sister chromatids break
- The centromeres of sister chromatids move apart, pulled by the microtubules of the spindle

Telophase

- Sister chromatids (now effectively separate chromosomes) reach opposite poles

Cytokinesis

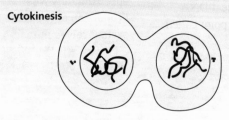

- The chromosomes decondense
- Nuclear envelopes appear around the chromosomes at each pole
- The spindle disappears
- The cell divides into two cells, by infolding of the plasma membrane in animal cells, or by formation of a new cell wall and plasma membrane in plants

Figure 46 Mitosis and cytokinesis

Knowledge check 22

An extract from crocuses, called colchicine, inhibits spindle formation. Suggest what effect colchicine will have on mitosis.

Core practical 3

Making a root tip squash

A good place to find cells going through the cell cycle is just behind the root tip of a young plant. To enable you to see the chromosomes, you need to:

- squash the root so that the cells are spread out in a single layer
- stain the DNA so that the chromosomes show up clearly when you observe the cells through a microscope. There are various ways of doing this, for example by using a red stain called acetic orcein, which makes DNA red.

Take a young root tip and cut off the end — you want just the 5 mm nearest to the tip. Put your root tip into a small glass container and cover it with acetic orcein stain, mixed with a little $1.0 \, mol \, dm^{-3}$ hydrochloric acid. Warm it gently for a few minutes. This will break the cells apart, making it easier to squash the root at the next stage. The stain will make the DNA look red.

Now put the stained root tip onto the centre of a microscope slide. Cut off the end 2 mm and throw this away. Put another couple of drops of acetic orcein onto the part still left on the slide.

Put a coverslip over the root tip on the slide, and wrap the central part of the slide in some filter paper. Gently, repeatedly tap on the filter paper above the coverslip with the blunt end of a pencil. You are trying to squash the root tip — without breaking the coverslip! Unwrap and check progress every now and then. Aim to get the root tip cells spread out as much as possible, but still in the same relative positions as they were to start with. You can add a bit more stain if it gets too dry.

When you are happy with your root tip squash, hold the slide over a Bunsen flame with your fingers for a few seconds — holding it with your fingers will stop you letting it get too hot. This is critical! It is very easy to overdo it, which will make the chromosomes disintegrate. Now you can look at the slide under the microscope. You should be able to see cells in various stages of mitosis.

Meiosis and gamete production

A few cells in the body — some of those in the testes and ovaries — are able to divide by another type of division, called **meiosis**. Meiosis involves two divisions (not one as in mitosis), so four daughter cells are formed. Meiosis produces four new cells with:

- only half the number of chromosomes as the parent cell
- different combinations of genes from each other and from the parent cell

In animals and flowering plants, meiosis produces gametes.

In a human, body cells are **diploid** ($2n$), containing two complete sets of chromosomes. Meiosis produces gametes that are **haploid** (n), containing one complete set of chromosomes.

Before meiosis begins, DNA replication takes place exactly as it does before mitosis. However, in the early stages of meiosis, homologous chromosomes (the two 'matching' chromosomes in a nucleus) pair up.

Each chromosome in a homologous pair carries genes for the same characteristics at the same locus. The alleles of the genes on the two chromosomes may be the same or different.

During meiosis, as the two homologous chromosomes lie side by side, their chromatids form links called **chiasmata** (singular: **chiasma**) with each other. When they move apart, a piece of chromatid from one chromosome can swap places with the other one. This is called **crossing over** (Figure 47). It results in each chromosome having different combinations of alleles from those it had before.

Knowledge check 23

Can mitosis occur in a haploid cell? Can meiosis occur in a haploid cell? Explain your answer.

 Homologous pair of chromosomes

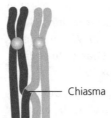 Chiasma

 Pieces of each chromatid have swapped places, resulting in new combinations of alleles

Knowledge check 24

Explain why crossing over between the two chromatids making up one chromosome would not have any effect.

Figure 47 Crossing over in meiosis

Another feature of meiosis that results in the shuffling of alleles — and therefore genetic variation — is **independent assortment**. During the first division of meiosis, the pairs of homologous chromosomes line up on the equator before being pulled to opposite ends of the cell. Each pair behaves independently from every other pair, so there are many different combinations that can end up together. Figure 48 shows the different combinations you can get with just two pairs of chromosomes. In a human, there are 23 pairs, so there are a huge number of different possibilities.

Chromosome mutation

Meiosis does not always proceed reliably. Alterations can occur in either the number or structure of the chromosomes that end up in the daughter cells.

Non-disjunction

Sometimes, the sets of chromosomes are not distributed correctly into the daughter cells. This is called **non-disjunction**. When the cell divides, one daughter cell may get two copies of a particular chromosome, while the other gets none.

Exam tip

Non-disjunction takes place during anaphase of meiosis 1.

For example, if the chromosome 21s fail to separate, a gamete may be formed that has two chromosome 21s, resulting in a zygote with three copies of this chromosome. This is an example of **polysomy**. It is the cause of **Down's syndrome**. The person usually has eyes that slant upwards, and may have health problems such as heart weakness. They have a naturally happy and friendly disposition.

If the sex chromosomes (XX or XY) fail to separate, then a gamete may be formed with no sex chromosome. If this fuses with a gamete containing an X chromosome, then a zygote with just one X chromosome and no Y chromosome is produced. This is an example of **monosomy**. This is the cause of **Turner's syndrome**. The person is female but may have non-functioning ovaries and sometimes physical and mental abnormalities.

Knowledge check 25

Which type of gamete could be the cause of Turner's syndrome — an egg, a sperm or either of them?

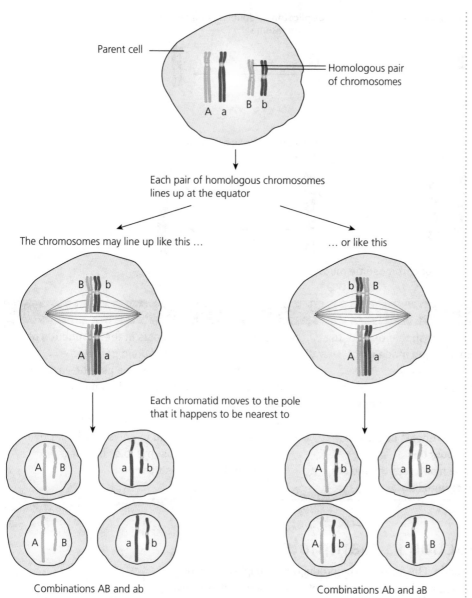

Figure 48 Independent assortment in meiosis

Knowledge check 26

Explain the difference between a pair of homologous chromosomes and a chromosome made up of chromatids.

Translocation

During chromosome separation in meiosis, one part of a chromosome may break off and attach itself to a different chromosome. This transfer is called **translocation** (Figure 49). Translocations might not cause any changes in phenotype, as long as no parts of the chromosomes have been lost. However, in some cases the presence of a particular length of DNA in the wrong place can have effects on other genes. For example, some types of cancer are caused by translocations, where the translocated DNA causes over-expression of other nearby genes, which can result in inappropriate proliferation of cells.

Deletion — loss of a section of chromosome

Duplicaton — doubling of a section of chromosome

Translocation — transfer of part of one chromosome to another chromosome

Inversion — a section of chromosome is inverted

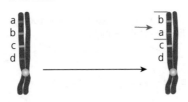

Figure 49 Translocation — a type of chromosome mutation

Summary

After studying this section, you should be able to:
- describe the cell cycle as being made up of interphase, mitosis and cytokinesis
- explain what happens to the genetic material during the cell cycle, including the stages of mitosis
- explain the importance of mitosis in growth, repair and asexual reproduction
- make a temporary squash of a root tip and stain it to observe the stages of mitosis
- explain the role of meiosis in the production of genetically dissimilar haploid gametes
- explain how meiosis produces genetic variation through crossing over and independent assortment
- describe chromosome mutations, including non-disjunction leading to polysomy (e.g. Down's syndrome) and monosomy (e.g. Turner's syndrome), and translocation

Sexual reproduction in mammals

All mammals reproduce sexually. This involves the production of haploid gametes, which fuse to form a diploid zygote. The zygote grows into an embryo and eventually an adult.

Gametogenesis

Gametogenesis is the production of gametes. In mammals, this occurs in the testes and ovaries. The production of spermatozoa or sperm is called spermatogenesis. The production of eggs is called oogenesis.

Spermatogenesis

Diploid **spermatogonia** at the edge of the seminiferous tubule (Figure 50) undergo mitosis and then meiosis to produce haploid **spermatids**. As meiosis proceeds, the cells move towards the centre of the tubule (Figure 51). The whole process takes about 64 days.

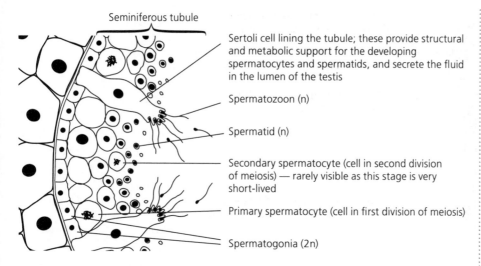

Sertoli cell lining the tubule; these provide structural and metabolic support for the developing spermatocytes and spermatids, and secrete the fluid in the lumen of the testis

Spermatozoon (n)

Spermatid (n)

Secondary spermatocyte (cell in second division of meiosis) — rarely visible as this stage is very short-lived

Primary spermatocyte (cell in first division of meiosis)

Spermatogonia (2n)

Seminiferous tubule

Figure 50 Histology of the testis

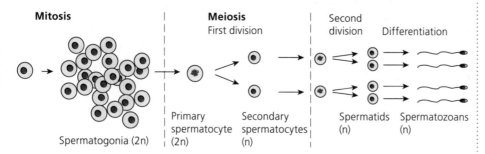

Figure 51 Sequence of spermatogenesis

Figure 52 shows the structure of a mature spermatozoon.

Oogenesis

Diploid **oogonia** divide by mitosis and then meiosis to produce haploid secondary oocytes. At birth, the oogonia have already become **primary oocytes**, in the first division of meiosis. The primary oocytes are inside primordial follicles, in which the oocyte is surrounded by a layer of granulosa cells (Figure 53). The primordial follicle remains in this state for many years.

At puberty, some of the primordial follicles develop into primary follicles, in which the primary oocyte grows larger and develops a coat called the **zona pellucida**. The primary follicle then develops into a secondary follicle, containing extra layers of granulosa cells, which are surrounded by a theca.

From puberty onwards, some of the primary follicles develop into **ovarian follicles**. These contain a large fluid-filled cavity, the antrum, between the granulosa cells. Just before ovulation, the primary oocyte completes the first division of meiosis to produce a **secondary oocyte** and a tiny **polar body** (Figures 54 and 55). At ovulation, the secondary oocyte is in metaphase of the second division of meiosis. After ovulation, the remains of the ovarian follicle develop into a corpus luteum.

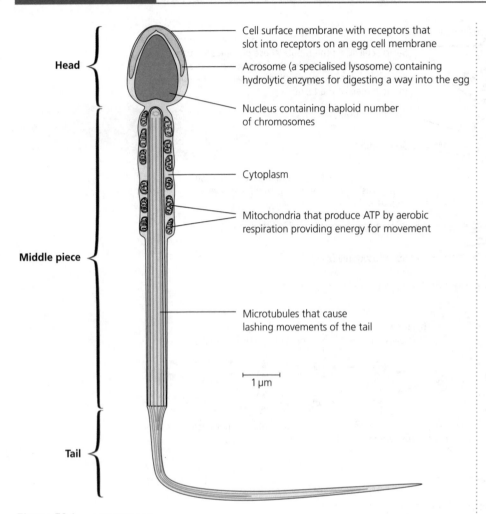

Head

Cell surface membrane with receptors that slot into receptors on an egg cell membrane

Acrosome (a specialised lysosome) containing hydrolytic enzymes for digesting a way into the egg

Nucleus containing haploid number of chromosomes

Cytoplasm

Mitochondria that produce ATP by aerobic respiration providing energy for movement

Middle piece

Microtubules that cause lashing movements of the tail

1 μm

Tail

Figure 52 A spermatozoon

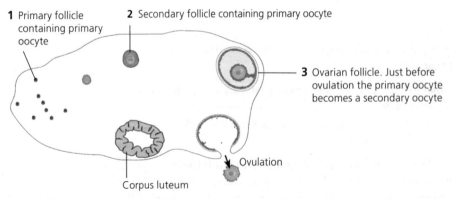

1 Primary follicle containing primary oocyte

2 Secondary follicle containing primary oocyte

3 Ovarian follicle. Just before ovulation the primary oocyte becomes a secondary oocyte

Corpus luteum

Ovulation

Figure 53 Histology of an ovary

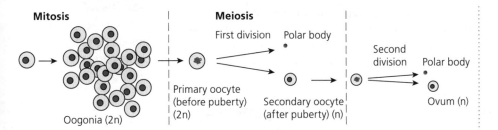

Figure 54 Sequence of oogenesis

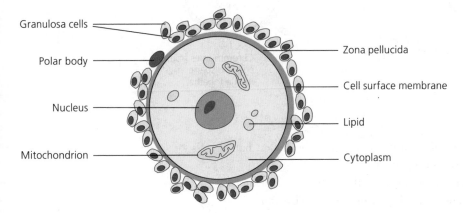

Figure 55 A secondary oocyte and polar body at ovulation

Fertilisation and development

Mammals have internal fertilisation. A liquid called semen, which contains sperm, is introduced into the vagina of a female mammal. The sperm swim towards the oviducts, where they may meet an egg.

The receptors on the cell surface membrane of the sperm cell lock into proteins in the cell surface membrane of the egg cell. This sets off a chain of events that results in the acrosome of the sperm cell releasing its hydrolytic enzymes. These digest the zona pellucida.

While this is happening, the egg responds to the binding of the sperm with its membrane by releasing specialised lysosomes called **cortical granules** into the zona pellucida. These cause changes in the protein molecules in the zona pellucida, altering its structure so that no more sperm can penetrate it. The altered zona pellucida is called a **fertilisation membrane**.

The chromosomes of the egg cell now complete meiosis. However, only one new complete cell is formed — the rest of the chromosomes, instead of forming nuclei in new cells, are 'discarded' and form a tiny, useless polar body. So the egg now has one complete set of chromosomes, just as the sperm cell does.

The chromosomes of the sperm join those of the egg. This is the point at which fertilisation happens — it is the fusion of the nuclei of the two gametes (Figure 56). The cell is now diploid, and it is called a zygote. This cell is how every young mammal begins. It divides over and over again to eventually form a complete organism.

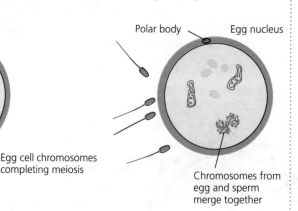

Figure 56 Fertilisation in a human

The diploid zygote divides repeatedly by mitosis to form a tiny, hollow ball of cells called a blastocyst (Figure 57). This is made up of:

- an inner cell mass, which will eventually form the embryo and the amnion
- a ring of cells called the trophoblast, which will form the placenta
- a fluid-filled cavity called the blastocoel

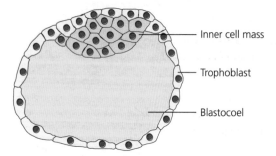

Figure 57 A blastocyst

Summary

After studying this section, you should be able to:
- describe how spermatogenesis and oogenesis take place
- describe the events of fertilisation from first contact between the gametes to the fusion of nuclei
- describe how the early embryo develops to the blastocyst stage

Sexual reproduction in plants

The flowers of a plant are its sexual reproduction organs (Figure 58). Many species of plant are hermaphrodite — one plant makes both male gametes and female gametes.

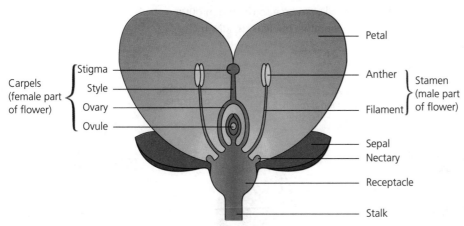

Figure 58 Structure of a flower

Formation of male and female gametes

In a flowering plant, the male gametes are found inside the **pollen grains** (Figure 59). Each pollen grain contains two haploid nuclei. One is called the **tube nucleus**. The other is the **generative nucleus**, and this will form the male gametes.

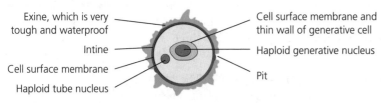

Figure 59 A pollen grain

The female gametes are inside the **ovules**, inside the ovaries. Each ovule contains a structure called an **embryo sac** (Figure 60). This contains six haploid cells and one diploid cell.

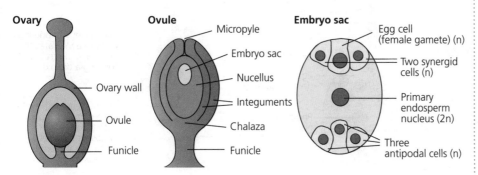

Figure 60 An ovary, an ovule and an embryo sac

> **Exam tip**
>
> It is incorrect to say that pollen grains 'are' male gametes. They contain male gametes.

Pollination and fertilisation

Pollen grains are carried from one flower to another by insects, birds or the wind. The pollen grain is left on the stigma of a flower. This is called pollination. If the stigma is the same species as the pollen grain, and if it is ripe, it secretes substances that stimulate the pollen grain to germinate (Figure 61). An area in the tough outer covering breaks down, and a tube grows out. The generative nucleus divides by mitosis to form two male gametes.

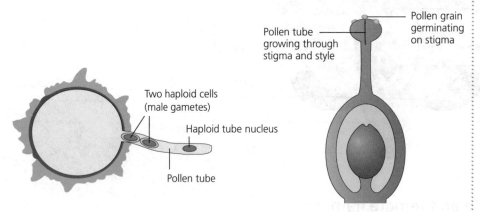

Figure 61 Germination of a pollen grain on a stigma

The pollen tube grows down through the style and enters the ovule through a little gap in its coverings called the **micropyle**. Once the tube has reached the embryo sac, the tube nucleus disintegrates. The two male gametes enter the embryo sac. One of them fuses with the egg cell, forming a diploid zygote. The other fuses with the diploid nucleus, forming a **triploid endosperm nucleus**. (A triploid cell has three complete sets of chromosomes.)

This strange double fertilisation is found only in flowering plants (Figure 62). The diploid zygote will develop into the embryo, eventually growing into a new plant. The triploid endosperm nucleus will divide to form a tissue called the **endosperm**, which provides nutrients for the young embryo during its early stages of germination. The whole ovule becomes a seed.

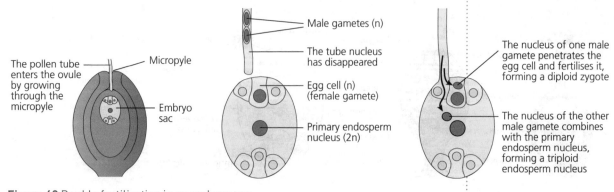

Figure 62 Double fertilisation in an embryo sac

Core practical 4

Investigating the effect of different sucrose concentrations on pollen tube growth

Pollen grains can be stimulated to grow tubes by exposing them to solutions containing chemicals similar to those that they encounter on a stigma, including sucrose. Pollen grains from different species respond to different concentrations of sucrose.

Make up sucrose solutions of different concentrations. Place drops into the centres of clean microscope slides. Dust a few pollen grains onto the surface of the sucrose solution drop. Do not cover. Observe under a microscope for the next 15 minutes and record the number of pollen grains that have begun to grow tubes.

You can find a full protocol for this investigation at www.nuffieldfoundation.org/practical-biology/observing-growth-pollen-tubes, from where you can download a student sheet.

Summary

After studying this topic, you should be able to:

- describe how pollen grains form in anthers, and embryo sacs form in ovules
- explain how the male nuclei form by division of the generative nucleus
- describe how the male nuclei reach the embryo sac, including the roles of the tube nucleus, pollen tube and enzymes
- describe how double fertilisation takes place inside an embyro sac to form a triploid endosperm and a zygote
- carry out an investigation to find the effect of varying sucrose concentrations on the growth of pollen tubes

Questions & Answers

In this section there are two sample examination papers, which contain questions in styles similar to those in the Edexcel AS GCE Paper 1 and Advanced GCE Papers 1 and 2 of the Biology B specification. However, whereas those papers test content from all or most of the topics you will study during your course, the sample papers here test only content from Topics 1 and 2.

You have 1 hour 30 minutes to do each paper. There are 80 marks on the paper, so you can spend 1 minute per mark, plus time for reading the questions and checking your answers. If you find you are spending too long on one question, move on to another that you can answer more quickly. If you have time at the end, come back to the difficult one.

Some of the questions require you to recall information that you have learned. Be guided by the number of marks awarded to suggest how much detail you should give in your answer. The more marks there are, the more information you need to give.

Some of the questions require you to use your knowledge and understanding in new situations. Don't be surprised to find something completely new in a question — something you have not seen before. Just think carefully about it, and find something that you do know that will help you to answer it.

Think carefully before you begin to write. The best answers are short and relevant — if you target your answer well, you can get many marks for a small amount of writing. Don't ramble on and say the same thing several times over, or wander off into answers that have nothing to do with the question. As a general rule, there will be twice as many answer lines as marks. So you should try to answer a 3-mark question in no more than six lines of writing. If you are writing much more than that, you almost certainly haven't focused your answer tightly enough.

Look carefully at exactly what each question wants you to do. For example, if it asks you to 'Explain', then you need to say how or why something happens, not just what happens. Many students lose large numbers of marks by not reading the question carefully.

Comments

Each question is followed by a brief analysis of what to watch out for when answering the question (shown by the icon ⓔ). All student responses are then followed by comments that indicate where credit is due. These are preceded by the icon ⓔ. In the weaker answers, they also point out areas for improvement, specific problems, and common errors such as lack of clarity, weak or non-existent development, irrelevance, misinterpretation of the question and mistaken meanings of terms.

Sample paper 1

Time allowed: 1 hour 30 minutes. Total marks available: 80

Question 1

Figure 1 shows five molecules found in living organisms.

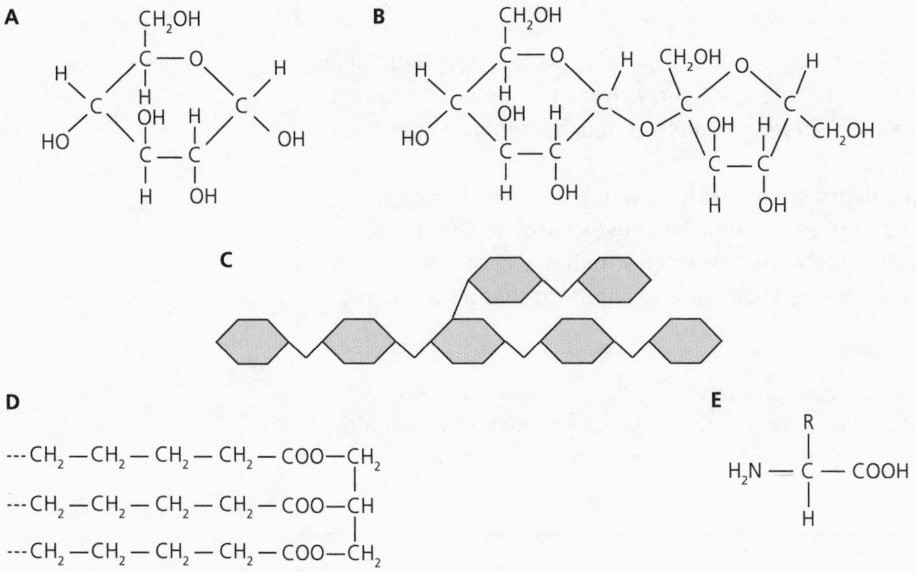

A

B

C

D

E

Figure 1

(a) Give the letter of one molecule that fits each of these descriptions:

  (i) the form in which carbohydrates are transported through phloem tissue in plants   (1 mark)

  (ii) the form in which carbohydrates are stored in animals   (1 mark)

  (iii) a molecule that is insoluble in water   (1 mark)

  (iv) a molecule that links together with others to form a polypeptide   (1 mark)

  (v) a molecule that contains ester bonds   (1 mark)

(b) Explain how the structure of water molecules makes water a good solvent.   (3 marks)

Total: 8 marks

e This question tests your ability to recall structures and functions of biological molecules. Note that you should only give *one* letter for each answer in (a). Part (b) ask you to 'Explain', which means that you need to say *how* the structure of the molecules results in water being a good solvent.

Student A

(a) (i) B ✓

(ii) C ✓

(iii) D or C ✓

(iv) E ✓

(v) D ✓

ⓔ **5/5 marks awarded** However, student A took an unnecessary risk with (a) (iii) by giving two answers. If the second one had been wrong, it could have negated the first correct one. If you are asked for one answer, it is best to give only one.

(b) In a water molecule, the hydrogen atoms have a tiny positive electrical charge and the oxygen atom has a similar negative charge ✓. Other atoms or ions with electrical charges ✓ are attracted ✓ to these charges on the water molecules. This makes them spread about ✓ among the water molecules.

ⓔ **3/3 marks awarded** This is a good answer — it explains how a substance dissolves in water and relates this clearly to the structure of a water molecule. Student A has made four possible marking points, but there is a maximum of 3 marks available.

Student B

(a) (i) A ✗

(ii) C ✓

(iii) E ✗

(iv) E ✓

(v) D ✓

ⓔ **3/5 marks awarded** In (a)(i) A is a glucose molecule, but plants transport sucrose. Even if you do not know what a sucrose molecule looks like, you should know that it is a disaccharide. In (a)(iii) amino acids are soluble. Either C or D would be correct.

(b) Water has dipoles and hydrogen bonds ✓, which help it to dissolve other substances.

ⓔ **1/3 marks awarded** There are no wrong statements in this answer, but it does not explain *why* water is a good solvent — it just states two facts about water molecules.

Question 2

Turner's syndrome is a condition caused by nondisjunction during meiosis 1.

(a) Which sex chromosomes are contained in the cells of a person with Turner's syndrome?

A   X only

B   XX

C   Y only

D   YX   (1 mark)

(b) At what stage of meiosis 1 does nondisjunction happen?

A   prophase

B   metaphase

C   anaphase

D   telophase   (1 mark)

(c) Turner's syndrome is the result of a chromosome mutation. Explain the difference between gene mutation and chromosome mutation.   (2 marks)

(d) The heights of two groups of girls with Turner's syndrome were measured from 1 year after birth until they were 19 years old. Group B girls were treated with oestrogen from the age of 15 onwards. Group A did not receive oestrogen treatment. The results are shown in Table 1.

Age/years	Group A (no oestrogen treatment)		Group B (treated with oestrogen from age 15)	
	Mean height/cm	Standard deviation	Mean height/cm	Standard deviation
1	63.8	6.8	64.9	3.7
3	84.2	4.7	83.1	4.4
5	96.5	5.0	92.7	5.4
7	106.0	5.0	102.3	5.1
9	115.0	5.7	112.5	4.9
11	122.8	6.2	119.8	4.6
13	130.7	5.8	128.1	6.0
15	135.8	5.7	134.2	5.9
17	140.6	6.0	139.1	5.2
19	143.2	6.1	143.1	3.7

Table 1

(i) The mean rate of growth for the girls in Group A between the ages of 1 and 19 is $4.18 \, cm \, year^{-1}$. Calculate the mean rate of growth for the girls in Group B between the ages of 1 and 19. Show your working.   (2 marks)

(ii) Explain what information is provided by the figures for standard deviation.   (2 marks)

(iii) Discuss whether these data indicate that treatment with oestrogen affects growth rate, and suggest what other information would be required in order to be able to draw a firm conclusion. (5 marks)

Total: 13 marks

ⓔ The first part of this question tests your knowledge, but the second part requires you to think carefully about a set of data, and to consider whether or not they provide any firm evidence of the effect of a particular treatment. It is always important to use clear and precise language, and this is especially so for part (c).

Student A

(a) A ✓

ⓔ **1/1 mark awarded** The person is female, but with only one, rather than the usual two, X chromosomes.

(b) B ✗

ⓔ **0/1 mark awarded** 'Disjunction' means the separation of the chromosomes, and this happens during anaphase.

(c) A gene mutation is when something goes wrong with a gene, and a chromosome mutation is when there is a mistake in a chromosome.

ⓔ **0/2 marks awarded** This is a weak answer. It does not really go any further than the wording already given in the question, and fails to explain what either a gene mutation or a chromosome mutation is.

(d) (i) growth = 143.1 − 64.9 = 78.2
So:
mean rate of growth = $\frac{78.2}{19}$ = 4.1157894 cm per year ✓✗

(ii) The standard deviation tells you how much variation there is in the figures. ✗

(iii) The growth rate in Group B is only 4.11, and the growth rate in Group A is 4.18, so it looks as though giving oestrogen actually makes the growth rate less. To find out if this is true, you would need to do the experiment again.

ⓔ **1/9 marks awarded** In part (i) the calculation is correct, but student A has failed to give the answer to one decimal place, to match the figures given in the table. Part (ii) has the beginnings of the right idea, but it is not quite precise enough to gain the mark. Standard deviation is an indication of the spread of the data on either side of the mean. In part (iii) student A has quoted the mean growth rates for both groups, but has not made clear that these are the mean growth rates over the whole 19 years of the study. Oestrogen was only given from the age of 15 onwards, so these figures are not the best or only comparisons that should be made. There is not enough in this answer to gain any marks at all.

Student B

(a) A ✓

e **1/1 mark awarded**

(b) C ✓

e **1/1 mark awarded**

(c) A gene mutation happens when there is a change in the base sequence of the DNA coding for a polypeptide ✓. This can sometimes change the sequence of amino acids in the polypeptide. A chromosome mutation happens when a piece of one chromosome breaks off and gets attached to a different one ✓.

e **2/2 marks awarded** Both gene mutation and chromosome mutation are correctly explained, making the essential difference between them clear.

(d) (i) change in mean height = 143.1 − 64.9 = 78.2

mean rate of growth = $\dfrac{78.2}{19}$ = 4.12 cm per year ✓✓

(ii) If you had all the individual heights at one age, and you drew a graph of number of girls at each height, you would get a normal distribution curve. The standard deviation tells you how much the individual heights spread on either side of the mean height ✓✓.

(iii) The girls in Group A were not given any oestrogen, but the girls in Group B were given oestrogen when they were 15. It might have made a little bit of a difference, because Group B grew 4.9 cm between 15 and 17 years, while Group A only grew 4.8 cm ✓ but that is only a small difference so it is probably not significant ✓.

On the other hand, the Group B girls were actually 1.6 cm shorter than the Group A girls at age 15, but by the time they were 19 they were only 0.1 cm shorter ✓, so maybe it was the oestrogen that made the difference and helped them to catch up ✓. But we can't say for sure, because we are not told how many girls there are in each group ✓ — you'd need to have large numbers before you could draw a conclusion. We also don't know what else might have been different between the two groups ✓, say how much they ate or how much exercise they took, so we would need to know whether other things were kept constant.

e **9/9 marks awarded** Part (i) is correct and a good, clear answer is given for part (ii). For part (iii) student B has looked very hard at the figures and made several good points, concentrating on the growth rate after oestrogen was given at age 15, and has calculated differences between the two groups immediately after this, and also by the end of the study at age 19. They have identified the importance of having large numbers in a study like this, and needing to know about other variables. This is a good answer.

Question 3

(a) Figure 2 shows a prokaryotic cell and a plant cell from a leaf.

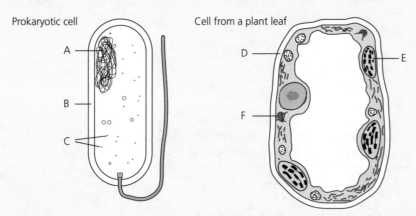

Prokaryotic cell   Cell from a plant leaf

Figure 2

    (i) Name the structures labelled A to F.   (6 marks)

    (ii) State two features of the prokaryotic cell that differ from all eukaryotic cells.   (2 marks)

    (iii) State one difference between structures B and D.   (1 mark)

    (iv) Give the letter of the structure that is likely to contain the most magnesium, and outline the role of magnesium in the cell.   (2 marks)

(b) Electron microscopes provide more information about the ultrastructure of cells because they have greater resolution than light microscopes. Explain the difference between resolution and magnification.   (2 marks)

Total: 13 marks

ⓔ Part (a) is straightforward — you just need to take care to get names and descriptions absolutely correct. Read part (b) carefully, then jot down at least three ideas that you can include, before starting to write your answer.

Student A

(a) (i) A Chromosome ✓
      B Cell wall ✓
      C Ribosome ✓
      D Cellulose cell wall ✓
      E Chlorophyll ✗
      F Golgi apparatus ✓

    (ii) It does not have a nucleus ✓. It does not have cellulose in its cell walls.

    (iii) The cell wall in the bacterium is not made of cellulose ✓.

    (iv) E ✓ because chlorophyll ✓ contains magnesium.

ⓔ **9/11 marks awarded** In part (i) E should be chloroplast. In part (ii) not all eukaryotic cells have cell walls, so although this second statement is correct it is not an appropriate answer to the question. Parts (iii) and (iv) are both correct.

> **(b)** Resolution means how much detail you can see, and magnification means how much bigger something is ✓.

ⓔ 1/2 marks awarded This answer suggests that student A does understand the meaning of each term, but needed to be a little more precise in order to gain both marks.

Student B

(a) (i) A Loop of DNA ✓
B Cell wall ✓
C Ribosome ✓
D Cell wall ✓
E Chloroplast ✓
F Golgi apparatus ✓

(ii) It has smaller ribosomes ✓ and no nuclear envelope ✓.

(iii) The cell wall of a bacterium is made of peptidoglycans ✓, whereas the cell wall of a plant cell is made of cellulose.

(iv) The nucleus, because magnesium is part of spindle fibres ✗✗.

ⓔ 9/11 marks awarded Parts (i), (ii) and (iii) are all correct. The main role of magnesium in a plant cell is as part of chlorophyll molecules, and these are found in the chloroplasts.

> **(b)** Resolution is the ability to distinguish two small objects — the smaller the objects you can distinguish, the greater the resolution ✓. Magnification is how many times greater the image is than the object ✓. You can keep on increasing magnification, but if the resolution stays the same you cannot see any more detail.

ⓔ 2/2 marks awarded This clearly explains the meanings of both words. The last sentence is not really necessary, but it does help to show that student B understands the difference between these two terms.

Question 4

Human blood contains a soluble, globular protein called fibrinogen. Fibrinogen is made up of six polypeptide chains, held together by disulfide bridges.

When a blood vessel is cut open, an enzyme called thrombin converts fibrinogen molecules to a fibrous protein, fibrin. Fibrin is insoluble and forms fibres that trap blood cells and begin to form a blood clot.

(a) (i) Suggest why fibrinogen is soluble in water but fibrin is not. (3 marks)

(ii) With reference to fibrinogen, explain the meaning of the term *quaternary structure*. (2 marks)

(iii) Thrombin is unable to act on other proteins found in blood plasma, such as albumin. Explain why. (2 marks)

(b) Hirudin is a substance obtained from leeches that acts as an inhibitor of thrombin. Hirudin is sometimes used in medicine to reduce the chance of formation of blood clots in people with disorders of the circulatory system.

An investigation was carried out to find out how the concentration of hirudin affects the rate at which blood clots. All other factors, including the concentrations of fibrinogen and thrombin, were kept constant. The results are shown in Figure 3.

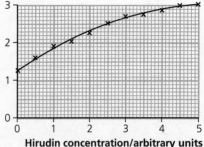

Figure 3

(i) Describe the effect of hirudin on clotting time. (2 marks)

(ii) With reference to Figure 3, suggest whether it is possible to conclude whether hirudin acts as a competitive or non-competitive inhibitor of thrombin. Explain your answer. (2 marks)

Total: 11 marks

 It is unlikely that you will know much about fibrinogen and fibrin, so read the information you are provided with carefully, and make use of it when answering the questions. Part (b)(ii) is quite tricky; don't be afraid to state that a conclusion is impossible if you believe this to be the case, but make sure that you support your answer by explaining the reason for your decision.

Student A

(a) (i) Fibrinogen might be hydrophilic and fibrin might be hydrophobic ✓.

(ii) This means the protein is made up of several polypeptide chains linked together ✓.

(iii) Thrombin is an enzyme, and enzymes are specific ✓ because their active site is only the right shape for one kind of substrate ✓.

 4/7 marks awarded The answer to part (i) is awarded 1 mark for the use of the terms hydrophobic and hydrophilic, but to get full credit there needs to be some explanation of what this means and also how the structure of the proteins might cause this difference in properties. Student A has not made use of the information in the question about fibrin forming fibres, which suggests that it is a fibrous protein. Part (ii) is correct, but it does not refer to fibrin as required, so only 1 mark is awarded. There is just enough detail in part (iii) for both marks.

(b) (i) The more hirudin there is, the longer the clotting time ✓.

(ii) I think it is a competitive inhibitor because it stops the substrate getting into the active site ✗.

ⓔ **1/4 marks awarded** In part (i) the general trend is correctly described, but there is no description of the shape of the curve, and no use is made of any figures. Student A has misunderstood what is required for part (ii). They should have explained how the graph does or does not provide evidence to suggest what type of inhibitor this is.

Student B

(a) (i) Fibrinogen is a globular protein ✓. It must have hydrophilic R groups on the outside of the molecule ✓ that can interact with the dipoles on water molecules ✓ and make it soluble. Fibrin is a fibrous protein. It is probably very long ✓ and big and so cannot dissolve.

(ii) This means that fibrinogen is made up of several polypeptide chains ✓. We are told there are six and they are linked by disulfide bonds ✓.

(iii) The active site of thrombin is complementary in shape ✓ to fibrinogen. Other proteins like albumin won't fit into its active site ✓ so it cannot affect them. Each enzyme is specific to its substrate.

ⓔ **7/7 marks awarded** There is good use of terminology in part (i) — the two proteins are identified as being globular and fibrous, and student B has also explained solubility in terms of hydrophilic groups and the dipoles on water molecules. Parts (ii) and (iii) are correct and expressed clearly.

(b) (i) The more hirudin there is, the longer it takes the blood to clot ✓. As the concentration of hirudin increases, the slope of the line decreases and starts to level off ✓.

(ii) To know whether it is a competitive inhibitor or not, we would need to find out what happens if the quantity of substrate is changed, because with a competitive inhibitor if you increase the concentration of the substrate in relation to the concentration of the inhibitor then the inhibition should be less ✓. With a non-competitive inhibitor the amount of substrate has no effect because the inhibitor is just binding with the enzyme whether there is any substrate there or not ✓. It is not possible to tell from this graph ✓, because with both competitive and non-competitive inhibitors you get less enzyme activity if you increase the inhibitor ✓.

ⓔ **4/4 marks awarded** Part (i) is a correct description of the general trend and also of the shape of the graph. In part (ii) student B shows that they know how to work out whether an inhibitor is competitive or non-competitive, and correctly states that there is no evidence here that we can use to make that decision.

Question 5

(a) Figure 4 shows the mitotic cell cycle.

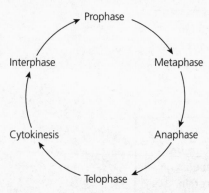

Figure 4

(i) On the diagram, write the letter A to indicate the stage at which DNA
replication occurs. (1 mark)

(ii) On the diagram, write the letter B to indicate the stage at which the
chromosomes line up on the equator of the spindle. (1 mark)

(b) Describe the events that occur during anaphase. (3 marks)

(c) A student made a root tip squash, stained it with acetic orcein and observed
it under a light microscope. She counted the number of cells she could see in
each stage of the cell cycle. Table 2 shows her results.

Stage of cell cycle	Number of cells observed
Prophase	12
Metaphase	10
Anaphase	4
Telophase	6
Cytokinesis	8
Interphase	160

Table 2

(i) Calculate the percentage of cells observed that were in metaphase. Show
your working. (2 marks)

(ii) Explain what these results show about the relative amount of time the
cells in this part of a plant spend in metaphase compared with interphase. (2 marks)

(iii) Describe how the student would prepare the root tip squash, for viewing
under a light microscope. (4 marks)

(iv) Describe how the student could investigate whether the temperature
at which the roots are kept affects the proportion of time spent in
metaphase compared with interphase. (You do not need to describe how
to prepare the root tip squash.) (6 marks)

Total: 19 marks

ⓔ Make sure you include at least four relevant, correct, clearly stated points for the 4-mark question part, and six for the 6-mark part. Remember to show your working in (c)(i).

Student A

(a)

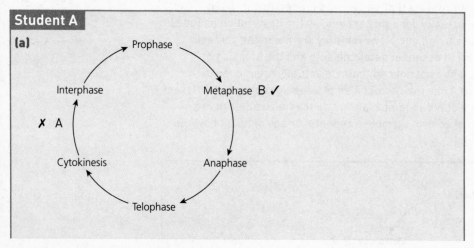

ⓔ **1/2 marks awarded** Part (i) is incorrect. The letter A needs to be clearly associated with interphase. The B for part (ii) is correctly positioned.

(b) The chromosomes separate into chromatids ✓. Each chromatid goes to the opposite end of the cell ✓. Then new nuclear membranes form and the chromatids disappear.

ⓔ **2/3 marks awarded** The first two sentences are relevant and correct, but do not give enough information. The last sentence is about telophase, not anaphase.

(c) (i) 20 ✗

(ii) The cells must spend longer in metaphase than in anaphase, because there are more ✓ of them in metaphase at any one time ✗.

(iii) Get a root tip and put it on a microscope slide. Put some filter paper round it and squash it with a pencil. Put some red stain onto it and warm it over a Bunsen flame ✓. Then look at it under a microscope.

(iv) Keep some root tips in a fridge and some others at room temperature in the lab. Measure both temperatures with a thermometer. Make and stain root tip squashes from both of them as described earlier. Then count the number of cells in metaphase and interphase in both sets and compare them.

ⓔ **3/14 marks awarded** The answer to part (i) is wrong and there is no working. In part (ii) student A has misread the question, which was about interphase not anaphase. However, 1 mark can be awarded for the idea that if there are more cells in a particular stage, this means that stage lasts longer. In part (iii) it appears that student A has actually done this activity, as the description includes several steps that would be carried out. However, they are not in the correct sequence, there is quite a lot of essential detail missing and the stain is not named — perhaps student A had not noticed that it is actually mentioned in the question. In part (iv) student A has the correct idea of keeping root tips at different temperatures, but there are only two temperatures suggested, rather than a range. There is no mention of control variables, repeats, or any detail of how the results will be collected.

Student B

(a)

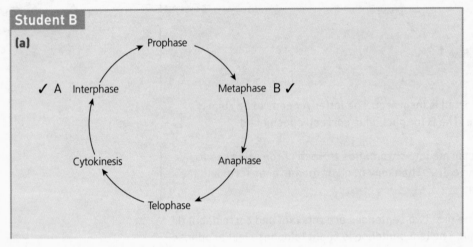

ⓔ **2/2 marks awarded**

(b) The spindle fibres, which are attached to the centromeres, pull ✓ them apart ✓. As the spindle fibres shorten ✓, they pull the chromatids ✓ to opposite ends ✓ of the cell.

ⓔ **3/3 marks awarded** This is an excellent answer. Student B mentions that the centromeres split and the chromatids are separated, and explains how the spindle fibres are involved in this process.

(c) (i) total number of cells counted = 200

percentage of cells in metaphase = $\dfrac{10}{200} \times 100$ ✓ = 5% ✓

(ii) There are 10 cells in metaphase and 160 in interphase, so interphase must last 16 times ✓ longer ✓ than metaphase.

(iii) Cut off a root tip. Put it on a slide and add some acetic orcein ✓. Hold the slide in your hand and warm it ✓ over a Bunsen flame. Put some more stain on it and cover it with a coverslip ✓. Wrap it in paper and tap with the blunt end of a pencil ✓ to squash the cells. Look at it under the microscope.

(iv) Temperature is the independent variable and the proportion of cells in the different stages is the dependent variable ✓.

Take at least 30 ✓ growing roots (for example, on a garlic bulb) and keep them at different temperatures — say 0, 10, 20, 30, 40, 50 and 60 °C ✓. Have three sets ✓ of roots at each temperature. Make sure all other conditions are exactly the same — quantity of water given, source of the water (so that it has the same nutrients) and light intensity ✓. The roots must all be from the same bulb of garlic, and they should all be the same age.

When all the roots have grown enough, cut them all off, taking care to label each one with its temperature. Cut off the end of each one, and prepare stained squashes from all of them, as explained in (iii) above. Make sure that each one is treated in the same way, for example given exactly the same amount of stain and kept in the stain for the same length of time. Place the squash under the microscope, and identify the stage of division in a sample of 100 cells ✓. Calculate the mean numbers for each set of three roots ✓. Plot a graph with temperature on the x-axis and percentage of cells in a particular stage on the y-axis ✓. Draw two lines — one for the percentage of cells in metaphase, and one for the percentage of cells in interphase.

ⓔ **14/14 marks awarded** The working and answer are correct in part (i). Part (ii) is entirely correct, and includes a calculation of how much longer interphase lasts. Part (iii) gets the maximum 4 marks, even though some detail is missing — for example, how much of the root tip to use and adding acid with the stain to help break the cells apart. Part (iv) is an excellent answer. Student B has begun by clarifying the variables, and has then suggested the range and the interval for the values of the independent variable, and described how other variables will be controlled. There is detail of how to collect results and how to process them. Repeats are included, and these are used to calculate mean values for each temperature. Perhaps there could have been some mention of how the different temperatures would be achieved, and also of how the results could be further processed to find the proportion of time spent in metaphase compared with interphase. However, maximum marks have been achieved.

Question 6

Figure 5 shows a small part of a DNA molecule during replication.

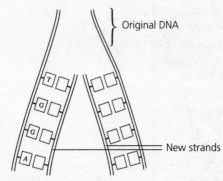

Figure 5

(a) (i) Name the part of a cell in which DNA replication takes place. (1 mark)

    (ii) Complete Figure 5 by writing the letters of the correct bases in each of the empty squares. (2 marks)

    (iii) Explain why this is known as semi-conservative replication. (2 marks)

(b) Meselson and Stahl carried out an experiment to determine whether DNA replication is conservative or semi-conservative. They grew bacteria for many generations in a medium containing heavy nitrogen (^{15}N) and then transferred them to a medium containing normal nitrogen (^{14}N).

DNA was extracted from the bacteria when they had been growing in the medium containing ^{15}N, and then when they had been growing for one and two generations in the medium containing ^{14}N.

The DNA samples were centrifuged in a calcium chloride solution. The heavier the DNA, the lower it came to rest in the centrifuge tube. Figure 6 shows some of the results.

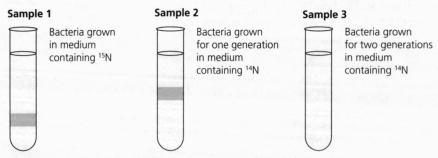

Figure 6

(i) Explain how the results for Sample 1 and Sample 2 support the hypothesis that DNA replication is semi-conservative and not conservative. (2 marks)

(ii) Complete Figure 6 by drawing the results you would expect in the tube for Sample 3. (1 mark)

Total: 8 marks

ⓔ There is a lot to read in this question. You will probably not know anything about Meselson and Stahl's experiment, so take time to read all the information thoroughly, and think carefully about the information that you are given in the question.

Student A

(a) (i) Nucleus ✓

(ii)

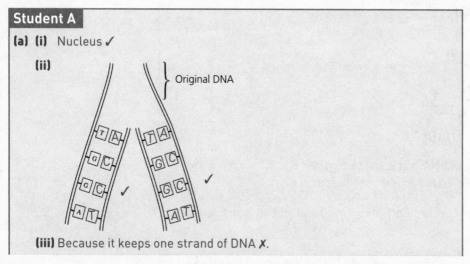

Original DNA

(iii) Because it keeps one strand of DNA ✗.

ⓔ **3/5 marks awarded** The answer to part (iii) is not clear enough to be awarded any marks. It is not obvious what 'it' means — student A is probably referring to the new DNA molecule but does not say so.

(b) (i) Because there is only one kind of DNA ✓ — there would be two sorts if it was conservative.

(ii)

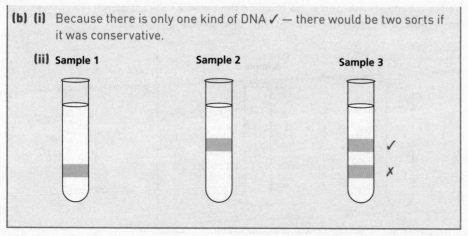

Sample 1   Sample 2   Sample 3

ⓔ **1/3 marks awarded** Student A is thinking along the right lines in part (i) and what is written is correct. However, for the second mark, the answer needs to say what the two kinds of DNA would be, or where they would appear in the tube, if the replication was conservative. In part (ii) you would expect to see two bands, one level with the band in sample 2 and one above the level of sample 2. The lower band is wrong — you would not get any all-heavy DNA because the cell is using ^{14}N to make its new DNA. So some of its new DNA will contain all ^{14}N, and some will contain half ^{14}N and half ^{15}N.

Student B

(a) (i) Nucleus ✓

(ii)

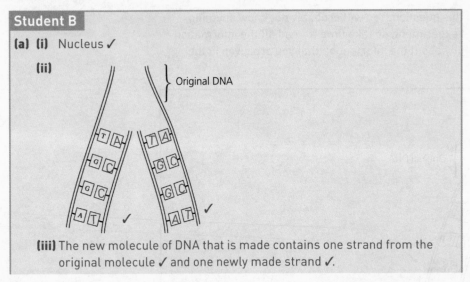

(iii) The new molecule of DNA that is made contains one strand from the original molecule ✓ and one newly made strand ✓.

e **5/5 marks awarded** Part (iii) is clearly expressed and all that is needed for full marks.

(b) (i) If it was conservative, then the new DNA molecules would contain all ^{14}N, so they would be light and would make a separate band higher up the tube ✓. The old DNA molecules would still contain all ^{15}N, so they would make a band low down in the tube ✓, like Sample 1. Having just one band for all the DNA means all the molecules must be the same ✓, with one strand in each molecule containing ^{14}N and the other strand containing ^{15}N ✓.

(ii)

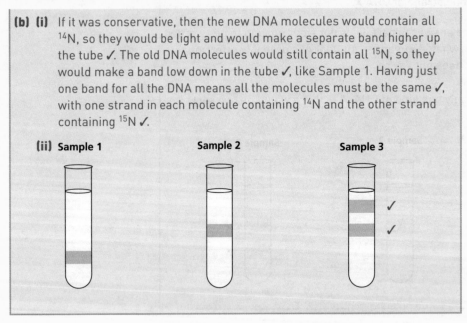

e **3/3 marks awarded** Part (i) is an excellent answer — though it would take up a lot more than the four lines allocated. Student B has described clearly what you would see if the type of replication was conservative (and why) and then done the same for what you would see if it was semi-conservative. Four ticks are shown, but a maximum of just 2 marks is available. Part (ii) is also correct.

Question 7

Figure 7 shows the stages in spermatogenesis in a mammal.

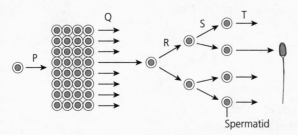

Figure 7

(a) State the letter(s) of the arrow or arrows that represent mitosis. (1 mark)

(b) Spermatogenesis in the nematode worm *Caenorhabditis elegans* follows a similar pattern to mammals. In both worms and mammals, spermatogenesis is sensitive to temperature. An investigation was carried out into the effects of mutations in two genes, A and B, that code for the production of two proteins involved in the control or spermatogenesis.

The results are shown in Table 3.

	Mean number of offspring produced	
	at 20°C	at 25°C
Normal males	280	150
Males with mutation in gene A	125	95
Males with mutation in gene B	220	85
Males with mutations in both gene A and gene B	90	0

Table 3

Compare the effect of increased temperature on the number of offspring produced by males with a mutation in gene A and males with a mutation in gene B. (3 marks)

(c) A study of the testes of the worms showed that:

- at 20°C, the number of spermatids in males with mutations in both gene A and gene B was the same as in normal males
- at 25°C, males with mutations in both gene A and gene B produced 29% fewer spermatids than normal males

Use this information to:

(i) suggest reasons for the difference in the number of offspring between normal males and males with mutations in both genes, at 20°C (2 marks)

(ii) give *one* letter from the diagram that could represent a stage of spermatogenesis that is affected by the mutations at 25°C but not at 20°C. Explain your answer. (2 marks)

Total: 8 marks

Questions & Answers

e You might like to begin this question by recalling what you know about spermatogenesis and jotting some labels onto the diagram to help you gather your thoughts. Remember that a good comparison of numerical data often involves manipulating the numbers in some way, for example by calculating a percentage difference or percentage change.

Student A

(a) P and T ✗

e **0/1 mark awarded** P is correct but T is not; T is the stage at which differentiation of spermatids to sperm occurs, with no further cell divisions.

(b) Temperature ✗ decreases the number of offspring for both groups. In the gene A mutation group it goes down from 125 to 95, a fall of 30, but in the gene B mutation group it goes down from 220 to 85, a fall of 135, which is more than four times more ✗.

e **0/3 marks awarded** The first sentence is unclear; student A needs to state that a rise in temperature (not just temperature) has this effect, or to state the two temperatures involved. They then make a comparison of the differences in the figures, but do not appreciate that because the number of offspring produced at 20 °C is much greater for the males with the mutation in gene B, then it is not useful to make a direct comparison between the size of the decrease in numbers when the temperature is raised to 25 °C. It would be much better to calculate a percentage decrease.

(c) (i) There was no difference in the number of spermatids made in the normal worms and the mutant worms, so it must have been something that went wrong after that ✓. Perhaps the spermatids didn't develop into sperm in the mutant worms ✓.

(ii) S ✓; because this is the step at which spermatids are formed from secondary spermatocytes ✓.

e **4/4 marks awarded** In part (i) student A has thought logically about the evidence provided, and has been able to make a suggestion that would explain the data. For part (ii) student A has worked out that the production of spermatids is affected at 25 °C and has correctly identified stage S as showing the production of spermatids from secondary spermatocytes.

Student B

(a) P ✓

e **1/1 mark awarded**

(b) The increase in temperature reduces the fertility of both types of worms ✓. In the ones with a mutation in gene A, the drop is 24% and in the ones with a mutation in gene B the drop is 61% ✓, so the effect is greater in the worms with a mutation in gene B ✓.

e **3/3 marks awarded** Student B clearly describes the effect of an increase in temperature on both types of worm, and then calculates the percentage change in each case, finishing with a conclusion based on the results of this calculation.

(c) (i) If both of them had the same number of spermatids, then it must have been a stage that happened later ✓ that caused the males with mutations in both genes to have less fertility than normal males. Perhaps the sperm were not well formed ✓ and could not swim well.

(ii) S ✓; because this time it was the number of spermatids that was different, and this is the stage where the spermatids are produced ✓.

e **4/4 marks awarded** Like student A, student B has been able to offer a good conclusion and a sensible suggestion to explain the data for part (i). Part (ii) is also correct, with a valid reason.

■Sample paper 2

Time allowed: 1 hour 30 minutes. Total marks available: 80

Question 1

A student carried out an investigation into the effect of substrate concentration on the rate of breakdown of hydrogen peroxide by the enzyme catalase.

He added catalase solution to samples of hydrogen peroxide of different concentrations and measured the volume of oxygen given off in the first 30 s of the reaction. Figure 1 shows his results.

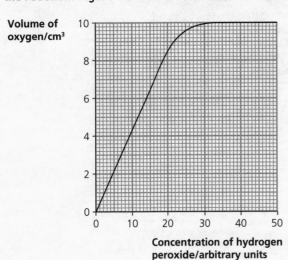

Figure 1

(a) State three variables that the student should keep constant in this investigation. (3 marks)

(b) Describe the results shown in Figure 1. (2 marks)

(c) Explain why it is important to measure the oxygen given off in the first 30 s, rather than timing how long it takes for the reaction to finish. (3 marks)

(d) Figure 2 shows the activation energy required to cause a substrate to change into a product.

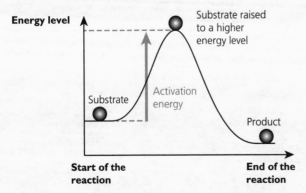

Figure 2

Sketch a similar diagram for the same reaction when catalysed by an enzyme. Label your diagram to show how it differs from Figure 2.

(2 marks)

Total: 10 marks

ⓔ Part (c) is not easy — try writing a list of points you want to make before constructing your final answer.

Student A

(a) enzyme concentration ✓, temperature ✓, volume

ⓔ **2/3 marks awarded** The first two are correct. 'Volume' is not enough — it could be volume of enzyme solution, volume of hydrogen peroxide solution or total volume. The answer needs to say what the volume is of.

(b) As the concentration of hydrogen peroxide increases, the volume of oxygen increases ✓ then levels off at about 10% hydrogen peroxide and 30 cm^3 of oxygen ✓.

ⓔ **2/2 marks awarded**

(c) Because this is when the reaction is fastest ✓.

ⓔ **1/3 marks awarded** This is true, but needs more for the second and third marks.

(d)

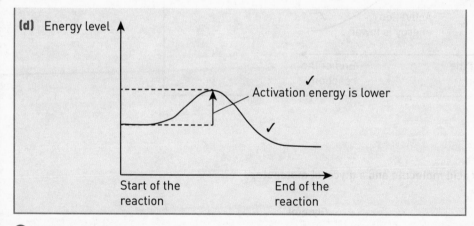

ⓔ **2/2 marks awarded**

Student B

(a) The length of time the oxygen is collected for ✓, the concentration of the enzyme ✓ and the temperature ✓.

ⓔ **3/3 marks awarded**

(b) As the substrate (hydrogen peroxide) increases, the rate of reaction (as measured by the volume of oxygen given off) also increases ✓, up until the hydrogen peroxide is 10%. Then it levels off.

ⓔ **1/2 marks awarded** This time, student B has not given quite such a good answer as student A, because they have not stated the value for the maximum oxygen volume.

(c) The reaction rate is fastest right at the start ✓, then it starts to slow down as the substrate gets changed into product ✓. So the substrate concentration varies with time — it doesn't stay constant ✓. If we measured how long it took for the reaction to finish, we'd actually find it took longer the more substrate there is ✓, because there would be more of it to break down.

ⓔ **3/3 marks awarded** This clear explanation shows that student B understands this difficult concept.

(d)

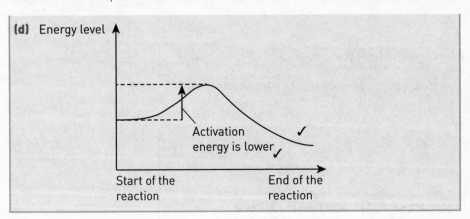

ⓔ **2/2 marks awarded**

Question 2

(a) Figure 3 shows a fatty acid molecule and a glycerol molecule.

Fatty acid

Glycerol

Figure 3

   (i) Draw a diagram to show how a fatty acid can form an ester bond with glycerol.

(3 marks)

(ii) What is the name for this type of reaction?

    A condensation

    B hydrolysis

    C neutralisation

    C polymerisation (1 mark)

(b) In many organisms, lipids function in waterproofing. For example, plant leaves and insect bodies are covered with a cuticle made of lipids.

Explain how the structure of lipid molecules relates to their roles as waterproofing substances. (2 marks)

(c) The grasshopper *Melanoplus sanguinipes* is found in California, USA. Figure 4 shows the rate of water loss from grasshoppers with different body masses, at 25 °C and at 42 °C.

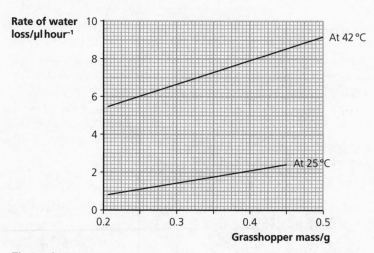

Figure 4

(i) Describe the effect of body mass on the rate of water loss at 25 °C. (2 marks)

(ii) Calculate the percentage of its body mass that is lost per hour at 25 °C by a grasshopper of mass 0.3 g. Show your working. (1 μl of water has a mass of 1 mg.) (3 marks)

(iii) Explain why the rate of water loss at all body masses is greater at 42 °C than at 25 °C. (2 marks)

(iv) With reference to the properties of water, explain what effect the evaporation of water from its body surface will have on the temperature of the grasshopper. (2 marks)

(v) The melting points of the cuticular lipids in two populations of *M. sanguinipes*, P and Q, were measured, and found to be 43 °C and 49 °C respectively. Population P lives in a cooler climate than population Q. Suggest how the higher melting point of the cuticular lipids of population Q could be an adaptation to the habitat. (2 marks)

Total: 17 marks

Questions & Answers

(a) (i)

H—C—C—C—C—C—C—C—C—O—C—OH ✗

(with H atoms above and below each carbon, a double-bonded O marked ✗, and branching H—C—OH, H—C—OH, H—C—OH, H groups) ✗

(ii) A ✓

e 1/4 marks awarded Part (i) is not correct. Student A has tried to show how the fatty acid would link to an oxygen on the glycerol molecule, rather than an OH group. Moreover, they have not shown that a water molecule is given off during this reaction.

(b) Lipids are hydrophobic ✓ and don't associate with water molecules, so they won't let them pass through.

e 1/2 marks awarded This does not answer the question, because it says nothing about the *structure* of lipid molecules. 1 mark has perhaps rather generously been given for the use of the term 'hydrophobic'.

(c) (i) As body mass increases the rate of water loss also increases. For a grasshopper that weights 0.2 g the rate of water loss is about 0.75 μl hour^{-1} and this increases to 2.4 μl hour^{-1} for a grasshopper that weighs 0.48 g. ✓

(ii) rate of water loss for a 0.3 g grasshopper = 1.4 μl hour^{-1}
So:

$$\text{percentage of body mass lost per hour} = \frac{1.4 \times 0.3}{100} \; ✓ = 0.0042 \, μl \, hour^{-1}$$

(iii) Grasshoppers move faster at higher temperatures, so they lose more water.

(iv) When water evaporates, it takes heat away with it, so it will cool the grasshopper down ✓.

(v) The grasshoppers in the hot place might melt if they had lipids with lower melting points, so they have to have lipids with high melting points.

ℯ 3/11 marks awarded The answer to part (i) correctly states the general trend in the relationship, but it is quite limited. It does not, for example, say that the relationship is linear, and (although it quotes two data points) it does not include any calculations, for example the increase in water loss for every extra gram of body mass. The answer to part (ii) gets 1 mark for correctly reading the value from the graph, but student A has forgotten how to calculate a percentage. In part (iii) it is probably correct that grasshoppers are more active at higher temperatures, but this does not explain the additional water loss fully. Part (iv) misses the chance to use the correct technical term — latent heat of evaporation. Part (v) is not correct. It is not the grasshoppers that would melt, only the lipids that cover their bodies.

Student B

(a) (i)

(ii) A ✓

ℯ 4/4 marks awarded Part (i) is entirely correct. Student B has shown the bond correctly and named it, and has also shown the water molecule that is given off during the reaction.

(b) Lipid molecules are mostly made up of long C–H chains ✓, which do not have OH groups or any other groups with charges on them ✓, so they are not attracted to the δ^+ and δ^- charges on water molecules ✓. So the water molecules don't get between the lipid molecules and separate them, which is what happens when something dissolves.

ℯ 2/2 marks awarded This is a full and clear answer to the question.

(c) (i) As body mass increases the rate of water loss increases ✓ linearly ✓. For every 1 g increase in mass, the water loss increases by $1.14\,\mu l\,hour^{-1}$. ✓

(ii) rate of water loss for a 0.3 g grasshopper = $1.4\,\mu l\,hour^{-1}$ ✓
So:

$$\text{percentage of body mass lost per hour} = \frac{1.4 \times 10^{-3}}{0.3} \times 100\ ✓ = 0.47\%\ ✓$$

(iii) At higher temperatures water molecules move around faster ✓, so more of them have enough energy to be able to escape from the liquid ✓. It is the more energetic molecules that leave, so the temperature of the water that stays behind gets lower, so the grasshopper cools down.

> **(iv)** It is the more energetic molecules that leave ✓, so the temperature of the water that stays behind gets lower ✓, so the grasshopper cools down ✓.
>
> **(v)** The grasshoppers that live in the warmer climate are more likely to survive if they have lipids covering their body that don't melt in the high temperatures they are in ✓, because if the lipids melted they would just run off their bodies ✓ and not stop water evaporating from them any more ✓. So only the grasshoppers with high-melting-point lipids survive and reproduce.

ⓔ **11/11 marks awarded** The answers to parts (i) and (ii) are excellent. Part (iii) is a correct explanation, but it is a pity that student B did not refer to the high latent heat of evaporation of water. Nevertheless, there is enough for full marks. The second sentence is not an answer to the question; if student B had read ahead, they would have realised it is the answer to part (iv), where the answer is repeated, gaining full marks. Part (v) shows that the student understands the importance of the lipids to the grasshoppers. There is a hint of a description of natural selection in the last sentence, which is nice to see, but not necessary in an answer to this question.

Question 3

In an investigation into the lytic cycle, a solution containing bacteriophages (viruses that infect bacteria) was mixed with a culture of bacteria. Samples were taken from the culture at 5-minute intervals, and the number of viruses was measured.

The results are shown in Figure 5.

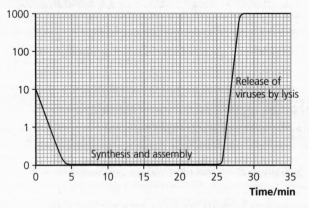

Figure 5

(a) Describe the events that took place between 0 and 5 minutes. (2 marks)

(b) Explain why no viruses were counted between 5 and 25 minutes. (2 marks)

(c) Assuming that all of the viruses released at the end of this cycle infected bacteria at 30 minutes, predict the number of viruses that would be counted at 60 minutes. (1 mark)

Total: 5 marks

ⓔ You might find it helpful to jot down what happens during the lytic cycle, and then annotate the graph to identify what is happening at each stage shown — this will help you with parts (a) and (b). Look carefully at the *y*-axis scale when you answer part (c).

Student A

(a) The number of viruses went down from 1 to nothing.

ⓔ **0/2 marks awarded** This is not an appropriate answer to the question, which requires the student to describe the events that took place, not the changes in the number of viruses.

(b) The viruses were all inside the bacteria ✓, so when the scientists looked at them through a microscope they could only see the bacteria and not the viruses.

ⓔ **1/2 marks awarded** The answer has been given credit for the idea that the viruses were inside the bacteria, but student A also needs to refer to the events of the lytic cycle to get full marks.

(c) 500 ✗

ⓔ **0/1 mark awarded** The answer is wrong; it looks as though it might be just a guess.

Student B

(a) This is when the virus was injecting its DNA into the bacteria ✓.

ⓔ **1/2 marks awarded** This is correct, but the answer should also include reference to the attachment of the viruses to receptors on the surface of the bacteria.

(b) There weren't any complete viruses ✓ during this time. New viruses are being made ✓ inside the bacteria, but they are just separate bits of DNA ✓ and proteins during this stage, and haven't been assembled into viruses yet.

ⓔ **2/2 marks awarded** This answers the question well. A fuller description of exactly what is happening at this stage of the viral lytic cycle could be given, but student A has done well in selecting the points that best answer the question.

(c) 100 000 ✓

ⓔ **1/1 mark awarded** In the first lytic cycle shown, 10 viruses becomes 1000, so in the second cycle we could expect those 1000 viruses to multiply 100 times to produce 100 000 viruses.

Question 4

(a) Figure 6 shows part of a flower just before fertilisation takes place.

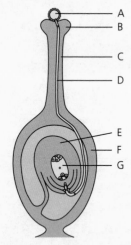

Figure 6

    (i)   Which letter labels the embryo sac?   (1 mark)

    (ii) On Figure 6, draw label lines to each of the following structures and label them.

        pollen tube nucleus   female gamete nucleus   (2 marks)

(b) Describe how fertilisation will occur in this flower.   (4 marks)

(c) Explain the importance of fertilisation in sexual reproduction.   (6 marks)

                                                 Total: 13 marks

e Note that (b) asks for a description, while (c) is an explanation. A description is simply *what* happens, whereas an explanation should include *why* or *how*.

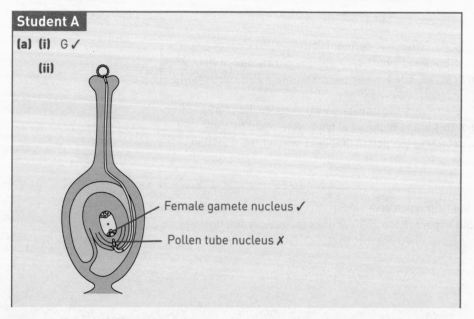

Student A

(a) (i) G ✓

    (ii)

ⓔ **2/3 marks awarded** For part (ii) student A has wrongly labelled one of the male nuclei instead of the tube nucleus, but the female gamete nucleus is correct.

> **(b)** The male nucleus will fuse with the female nucleus ✓ and this will make a zygote ✓, which will develop into an embryo.

ⓔ **2/4 marks awarded** Two correct statements score half the marks. The point about the embryo does not relate to the question, because it goes beyond the process of fertilisation. Student A has forgotten about the fusion of the other male nucleus with the endosperm nucleus.

> **(c)** Fertilisation means two haploid cells fuse to form a diploid cell ✓, so the new plant will have the right number of chromosomes in its cells. Fertilisation also produces variation.

ⓔ **1/6 marks awarded** This is not sufficiently detailed to gain more credit. The idea of maintenance of chromosome number from generation to generation is correct. However, the statement about variation is too vague to earn any marks.

Student B

(a) (i) G ✓

(ii)

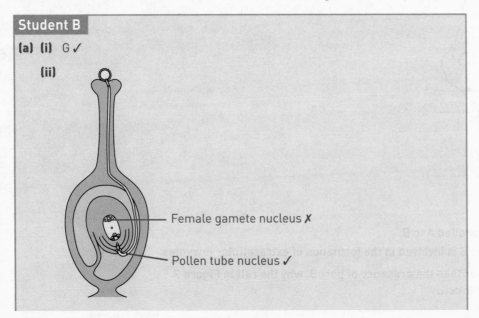

Female gamete nucleus ✗

Pollen tube nucleus ✓

ⓔ **2/3 marks awarded** In part (ii) the pollen tube nucleus is correct, but the female gamete nucleus is wrong — it is the one at the end closest to the micropyle (the gap in the integuments around the ovule).

> **(b)** One of the male nuclei fuses with the female nucleus ✓. They were both haploid so the zygote is diploid ✓. The other male nucleus fuses with the diploid nucleus ✓ in the middle of the embryo sac, producing a triploid nucleus ✓.

ⓔ **4/4 marks awarded** This is a concise and efficient answer.

(c) Sexual reproduction involves haploid gametes ✓. Fertilisation is the fusion of the nuclei of two gametes ✓. This forms a diploid zygote ✓. So fertilisation makes sure the new organism has the correct number of chromosomes ✓. It is also important in producing genetic variation ✓, as any male gamete can fuse with any female gamete, and because they will have different alleles ✓ the zygote can end up with different combinations of alleles from its parents ✓.

@ **6/6 marks awarded** This is an excellent answer.

Question 5

(a) Figure 7 shows a small part of a cell, as seen using an electron microscope.

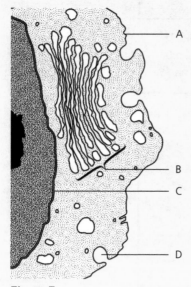

Figure 7

(i) Name the parts labelled A to D. (4 marks)

(ii) Describe how part B is involved in the formation of extracellular enzymes. (3 marks)

(b) Give *two* reasons, other than the presence of part B, why the cell in Figure 7 cannot be a prokaryotic cell. (2 marks)

Total: 9 marks

@ This is a straightforward question. Make sure you get the names in (a)(i) absolutely correct, and that you give *two* different reasons in (b).

Student A
(a) (i) **A** Plasma membrane ✓
B Golgi ✓
C Nucleus ✗
D Phagocyte ✗

(ii) First, the enzymes are made by protein synthesis on the ribosomes. Then they go into the endoplasmic reticulum. Then they are taken to the Golgi ✓ where they are packaged. Then they go in vesicles ✓ to the cell membrane where they are sent out by endocytosis.

ⓔ **4/7 marks awarded** In part (i) C is the nuclear envelope (or membrane), not the nucleus itself. A phagocyte is a cell — perhaps student A is thinking of a phagocytic vesicle. They have not thought properly about exactly what part (ii) was asking, and have wasted time writing about events that take place before and after the involvement of the Golgi apparatus. There is, however, a mark for the idea that the Golgi receives proteins that have been in the rER, and another for packaging them into vesicles. Note that the enzymes are secreted from the cell by *exocytosis*, not endocytosis.

(b) It has a nucleus ✓ and it has Golgi apparatus ✗.

ⓔ **1/2 marks awarded** The Golgi apparatus is part B, and this has been excluded by the question.

Student B

(a) (i)

    A Cell surface membrane ✓

    B Golgi apparatus ✓

    C Nuclear envelope ✓

    D Vesicle/endocytosis ✓

    (ii) Proteins made in the rER are transported to the convex face ✓ of the Golgi apparatus in vesicles. The vesicles fuse ✓ with the Golgi and the proteins inside are modified ✓ by adding sugars to make glycoproteins ✓. They are packaged inside membranes ✓ and sent to the cell membrane.

ⓔ **7/7 marks awarded** Part (ii) is an excellent and concise answer.

(b) If it was a prokaryotic cell it wouldn't have a nucleus ✓ and it would have a cell wall ✓.

ⓔ **2/2 marks awarded**

Question 6

Field beans, *Phaseolus vulgaris*, were grown in three different soils. The control plants grew in soil with all necessary mineral ions, while two other sets of plants were grown in soil that was deficient in either magnesium or phosphate.

After 10 weeks, the dry mass of the shoots and roots of each plant was recorded. The chlorophyll content of the leaves, measured as the mass of chlorophyll per g of plant material, was also recorded.

The results are shown in Table 1.

Treatment	Mean dry mass/g per plant		Ratio of shoot dry mass to root dry mass	Mean chlorophyll content/mg g^{-1} dry mass
	Shoots	Roots		
Control	2.50	0.50	5.0	11
Lacking Mg^{2+}	1.50	0.15	10.0	4
Lacking PO$_4^-$	0.90	0.48		12

Table 1

(a) What is the independent variable in this investigation?

    A   the chlorophyll content

    B   the dry mass

    C   the nutrients in the soil

    D   the ratio of shoot dry mass to root dry mass   (1 mark)

(b) Calculate the ratio of shoot dry mass to root dry mass for the beans lacking phosphate.   (1 mark)

(c) (i) Explain the results for chlorophyll content.   (3 marks)

    (ii) Suggest why plants with low chlorophyll content not only have a low dry mass of shoots, but also of roots.   (2 marks)

(d) (i) Compare the growth of the shoots and roots of the control plants with the plants that were lacking phosphate.   (2 marks)

    (ii) Explain why plants need phosphate.   (2 marks)

Total: 11 marks

@ Read parts (b), (c) and (d) carefully and make sure that you are looking at the right parts of the table. Remember to refer clearly to figures from the table where relevant, and do calculations if you are asked to compare figures.

Student A
(a) C ✓

@ **1/1 mark awarded**

(b) 0.9:0.48 = 1.875 ✓

@ **1/1 mark awarded** Although the mark has been given, it would have been better to round up the answer to the same number of decimal places as the original data.

(c) (i) Magnesium is needed for making chlorophyll ✓, so the plants lacking it only had a small amount of chlorophyll ✓. The ones lacking phosphate could still make chlorophyll ✓. They actually had a bit more than the control plants, but that difference is probably not significant.

(ii) Chlorophyll absorbs sunlight ✗ for photosynthesis ✓, which is needed to make food ✗ for the whole plant, including its roots.

ⓔ **4/5 marks awarded** In part (ii) It is not true to say that chlorophyll 'absorbs sunlight' — it absorbs energy from sunlight. Credit has been given for reference to photosynthesis, but making 'food' is poor use of terminology at this level.

(d) (i) The shoots of the plants lacking phosphate grew the least well of all the plants — only 0.9 dry mass compared with 2.5 for the control plants and 1.5 for the plants lacking magnesium. However, their roots grew almost as well as the control plants ✓. So their shoots were much more affected by the lack of phosphate than their roots were.

(ii) To make DNA ✓ and ATP ✓.

ⓔ **3/4 marks awarded** Part (i) does not quite answer the question. It should have focused on the differences between the control plants and those lacking phosphate, and there was no need to mention the plants lacking magnesium at all.

Student B

(a) C ✓

ⓔ **1/1 mark awarded**

(b) $\frac{0.9}{0.48} = 1.88$ ✓

ⓔ **1/1 mark awarded** Correct.

(c) (i) Chlorophyll molecules contain magnesium ✓, so with a shortage of magnesium less chlorophyll will be made. It is surprising that they made any at all, so perhaps the $4\,mg\,g^{-1}$ was chlorophyll that was already in the plants before they were put into the solution that didn't have any magnesium in it.

(ii) If they are lacking chlorophyll their rate of photosynthesis slows down ✓ and they cannot manufacture carbohydrates ✓. This means that they cannot make cellulose ✓ to make new cell walls ✓, so the whole of the plant suffers because it cannot make enough new cells to grow properly.

ⓔ **3/5 marks awarded** Part (i) begins well, but then loses its way a little. The second sentence shows that student B is thinking carefully, but in fact we were not told that the soil had absolutely no magnesium, just that it was deficient in it. No mention is made of the chlorophyll content of the other two groups of plants, so the answer is not complete. Part (ii) is an excellent answer, with plenty of relevant detail.

(d) (i) The shoots of the plants lacking phosphate had 1.6 g less dry mass ✓ than the control plants, but the roots had almost the same dry mass ✓. It looks as though phosphate has a big effect on shoot growth but does not affect root growth significantly ✓.

(ii) Phosphate is required for making ADP and ATP ✓, which is made from ADP during respiration ✓ and photosynthesis.

ⓔ **4/4 marks awarded** Part (i) is a brief answer, but well focused on the question. A calculation has been done, which is always a good idea when comparing figures.

Question 7

(a) Figure 8 shows a cell in various stages of the cell cycle.

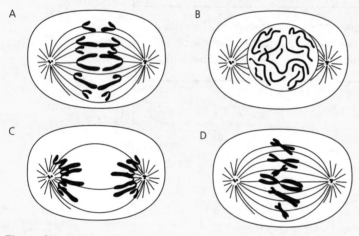

Figure 8

Name the stage represented by each diagram, and arrange them in the correct sequence.

(5 marks)

(b) Describe the role of spindle fibres (microtubules) in mitosis.

(3 marks)

(c) Figure 9 shows the changes in the mass of DNA and total cell mass during two cell cycles. Different vertical scales are used for the two lines.

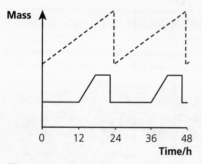

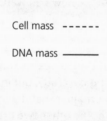

Cell mass - - - - - -

DNA mass ——————

Figure 9

   (i) On Figure 9, write the letter D to indicate a time at which DNA replication is taking place. (1 mark)

   (ii) On Figure 9, write the letter C to indicate a time at which cytokinesis is taking place. (1 mark)

(d) Mitosis produces new cells that are genetically identical to the parent cell. However, meiosis produces new cells that are genetically different.

   (i) Describe the roles of mitosis in living organisms. (2 marks)

   (ii) Outline the ways in which meiosis produces genetic variation. (3 marks)

Total: 15 marks

ⓔ The graph in (c) may be unfamiliar. If so, think carefully about what happens during the cell cycle. This should enable you to work out the answers.

Student A

(a) A metaphase ✓, B prophase ✓, C telophase ✓, D anaphase ✓

ⓔ **4/5 marks awarded** Student A has named each stage correctly, but has not arranged them in the right order.

(b) The spindle fibres pull the chromatids to opposite ends of the cell ✓.

ⓔ **1/3 marks awarded** This is correct, but there is not enough here for 3 marks.

(c)

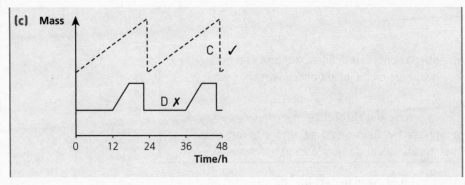

ⓔ **1/2 marks awarded** Cytokinesis is identified correctly, but DNA replication is not.

(d) (i) Mitosis is used in growth and repair ✓.

   (ii) Crossing over and independent assortment ✓.

ⓔ **2/5 marks awarded** Part (i) is correct, but not a good enough answer for 2 marks at AS. In part (ii) student A has identified the two processes that result in genetic variation in meiosis, but has not said anything about either of them, earning just 1 mark for the two names.

Student B

(a) B prophase ✓, A metaphase ✓, D anaphase ✓ C telophase ✓✓

ⓔ **5/5 marks awarded** All identified correctly, and in the right order.

(b) Spindle fibres are made by the centrioles. They latch onto the centromeres ✓ of the chromosomes and help them line up on the equator ✓. Then they pull ✓ on the centromeres so they come apart and they pull the chromatids ✓ to opposite ends of the cell in anaphase.

ⓔ **3/3 marks awarded** A good answer, with four correct points, but for a maximum 3 marks.

(c)

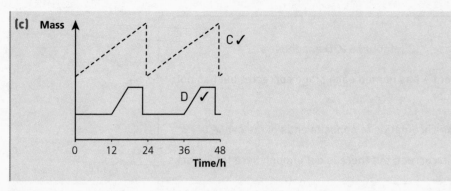

ⓔ **2/2 marks awarded**

(d) (i) Mitosis is used for growth. When a cell divides, the new cell gets exactly the same chromosomes as the original cell, which is what you want for growth ✓.

(ii) Independent assortment, where the chromosomes can line up any way on the equator in metaphase 1 ✓. Crossing over, where chromatids of homologous chromosomes break and rejoin during prophase ✓.

ⓔ **3/5 marks awarded** Only one point is made in part (i). The question has already stated that mitosis produces genetically identical cells, so there is no credit for stating this in the answer. Student B could have said more — for example, stem cells could have been mentioned in relation to growth. There is no mention of asexual reproduction, which is another important role of mitosis. The statement about independent assortment in part (ii) needs to make clear that it is the pairs of chromosomes that can 'line up any way'. Neither description says anything about alleles — for example, that during crossing over the alleles on one chromosome can be different from those on another, so that crossing over leads to different combinations of alleles on the chromosomes.

Knowledge check answers

1 In alpha glucose, the hydrogen on carbon 1 points upward, whereas in beta glucose it points downwards.

2 $C_5H_{10}O_5$

3 Carbon atoms 1 (on the left-hand molecule) and 4 (on the right-hand one).

4 Glycogen molecules are made of shorter chains of glucose molecules than starch molecules. Starch is actually a mixture of two substances — amylopectin and amylose — whereas glycogen is just a single substance. Amylose molecules curl up into a spiral, which glycogen does not do.

5 Both involve the removal of a water molecule, and so are condensation reactions.

6 Hydrolysis

7 They do not have any hydrophilic areas on their molecules, and so there are no parts of the molecule that will face outwards into water, like the heads of phospholipids.

8 The tertiary structure is held in shape by bonds (hydrogen, ionic and disulfide) that form between amino acids in the chain. These bonds form between particular R groups on the amino acids, and therefore the sequence of amino acids determines where the bonds can form, and therefore the three-dimensional shape of the folded chain.

9 Primary structure: peptide bonds. Secondary structure: hydrogen bonds. Tertiary and quaternary structure: hydrogen, ionic and disulfide bonds.

10 For example:

Haemoglobin	Collagen
Globular protein	Fibrous protein
Polypeptide chains of fixed length	Polypeptide chains not of fixed length
Primary structure not formed of repeating sections of the same amino acids	Primary structure formed of repeating sections of the same amino acids
Molecule made of four polypeptide chains	Molecule made up of three polypeptide chains
Soluble in water	Not soluble

11 DNA contains deoxyribose, while mRNA contains ribose. DNA contains the base thymine, while mRNA contains uracil. DNA is double-stranded, while mRNA is single-stranded.

12 The anticodon of three bases on a tRNA molecule determines which amino acid it will pick up. This anticodon then binds with its complementary codon on the mRNA molecule, ensuring that the correct amino acid is brought to add to the chain.

13 A triplet is a sequence of three bases on the sense strand of a DNA molecule, coding for one amino acid. A codon is a set of three bases on an mRNA molecule, and an anticodon is the complementary set of three bases on a tRNA molecule.

14 UGC CCA AUG GGC
UGC CAA UGG GC
UGC CCG AAU GGG C
UGC CGA AUG GGC

15 Hydrogen bonds help to maintain the secondary structure and tertiary structure of the enzyme molecule. If they are broken, the enzyme loses its shape, so the substrate can no longer fit into the active site and the reaction cannot be catalysed.

16 a All except pH

b Temperature, pH and non-competitive inhibitors

17 The scale bar with the diagram of the bacterium shows that 12 mm represents 0.1 μm. The length of the cell in the diagram is 55 mm. Therefore the length of the bacterium is $(55/12) \times 0.1$ μm = 0.46 μm. The scale bar with the diagram of the animal cell shows that 12 mm represents 10 μm. The length of the cell in the diagram is 100 mm (you may get a slightly different value, depending on exactly where you measure it). Therefore the length of the animal cell is $(100/12) \times 10$ μm = 83.3 μm.

The animal cell is therefore 83.3/0.46 = 181 times longer than the bacterium.

18

Prokaryotic cell	Eukaryotic cell
No nucleus or nuclear envelope	Nucleus surrounded by envelope (two membranes)
DNA in the form of a single circular molecule	DNA as several linear molecules, each forming a chromosome
Plasmids (small circular pieces of DNA) usually present	No plasmids
70 S ribosomes (20 nm diameter)	80 S ribosomes (30 nm diameter)
No endoplasmic reticulum or Golgi apparatus	Endoplasmic reticulum and Golgi apparatus present
Cell wall always present, made of cross-linked peptidoglycans	Cell wall present in plant cells and fungal cells; made of cellulose in plants and various substances in fungi
No cytoskeleton	Cytoskeleton of microtubules and microfilaments
No mitochondria or plastids	Mitochondria usually present; plastids (chloroplasts and amyloplasts) often present in plant cells

19 2.3×10^4 or $23\,000$

20 3.15×10^{-2} or 0.0315

21 The angle of the mitosis sector is about 10°, so it makes up about 1/36th or 2.8% of the total cycle.

22 The chromosomes will behave normally during prophase. However, they will not line up at the equator of the cell during metaphase, so the cell will become 'stuck' in prophase.

23 Mitosis can occur in a haploid cell. Meiosis cannot occur in a haploid cell, because the chromosomes will not be able to pair up.

24 The two chromatids are identical — they were formed by DNA replication. It will therefore make no difference if part of one is swapped with the other.

25 Either of them. Turner's syndrome will result if a sperm with no sex chromosome fuses with a normal egg carrying an X chromosome; it will also result if an egg with no sex chromosome fuses with a normal sperm carrying an X chromosome.

26 Each of a pair of homologous chromosomes is made up of two chromatids. The DNA sequences (alleles) on each chromosome may be different from one another. However, the two chromatids in a chromosome contain identical sequences of DNA.

Homologous chromosomes are not joined to each other (although they become temporarily joined when chiasmata form). The two chromatids of a chromosome are joined at the centromere.

27 There are many features that you could include in your table, for example:

Feature	Sperm cell	Egg cell	Explanation
Size	Very small	Relatively large	Egg cells are relatively large, as they contain food stores for the developing embryo before implantation. Sperm cells are relatively small, to reduce the mass that needs to be carried during movement, conserving energy.
Acrosome	Present	Not present	The acrosome contains enzymes that enable the head of the sperm cell to make a pathway into the egg cell.
Chromosome number	Haploid	Haploid	At fertilisation, the diploid number of chromosomes is restored.
State of cell division	Meiosis has been completed	Meiosis not completed	The egg completes meiosis when fertilisation takes place.
Mitochondria	Many, mostly in the tail	Few	Mitochondria produce ATP by aerobic respiration; this fuels the swimming movements of the sperm tail.
Cell surface membrane	Not folded	Folded to form microvilli	On the egg this aids binding of the sperm cell.
Lipid globules	Not present	Present	These provide energy to the embryo before implantation.

28 a 8

   b 16

   c 24

Index